WORLD BANK DISCUSSION PAPER NO. 330

China

Issues and Options in Greenhouse Gas Emissions Control

Report of a Joint Study Team from
The National Environmental Protection Agency of China,
The State Planning Commission of China,
United Nations Development Programme, and
The World Bank

Edited by
Todd M. Johnson,
Junfeng Li,
Zhongxiao Jiang,
and Robert P. Taylor

The World Bank
Washington, D.C.

Discussion Papers present results of country analysis or research that are circulated to encourage discussion and comment within the development community. To present these results with the least possible delay, the typescript of this paper has not been prepared in accordance with the procedures appropriate to formal printed texts, and the World Bank accepts no responsibility for errors. Some sources cited in this paper may be informal documents that are not readily available.

The complete backlist of publications from the World Bank is shown in the annual Index of Publications, which contains an alphabetical title list (with full ordering information) and indexes of subjects, authors, and countries and regions. The latest edition is available free of charge from the Distribution Unit, Office of the Publisher, The World Bank, 1818 H Street, N.W., Washington, D.C. 20433, U.S.A., or from Publications, The World Bank, 66, avenue d'Iéna, 75116 Paris, France.

ISSN: 0259-210X

Todd M. Johnson is an energy and environmental economist in the Environment and Municipal Development Division (Country Department II) of the World Bank's East Asia and Pacific Regional Office. Junfeng Li is an energy efficiency specialist at the Energy Research Institute, Beijing, China. Zhongxiao Jiang is a macroeconomist with the Chinese Academy of Social Sciences, Beijing. Robert P. Taylor is senior energy economist in the Infrastructure Division (Country Department II) of the Bank's East Asia and Pacific Regional Office.

Library of Congress Cataloging-in-Publication Data

China : issues and options in greenhouse gas emissions control /
    edited by Todd M. Johnson ... [et al.] ; report of a joint study
    team from the National Environmental Protection Agency of China ...
    [et al.].
        p.   cm. — (World Bank discussion papers ; 330)
    Includes bibliographical references.
    ISBN 0-8213-3660-6
    1. Greenhouse gases—Environmental aspects.   2. Energy
consumption—Environmental aspects—China.   3. Air quality
management—Economic aspects—China.   4. Air quality management—
China.   I. Johnson, Todd.   II. China.   Kuo chia huan ching pao hu
chü.   III. Series.
TD885.5.G73C474   1996
363.73′92′0951—dc20
                                                96-20971
                                                   CIP

Contents

Foreword

Global climate change is one of the looming environmental threats of the 21st century. Although the speed and magnitude of the so-called "greenhouse effect" is not yet known, scientific evidence is mounting that the increase in worldwide atmospheric concentrations of CO_2 and other long-lived greenhouse gases that has been accelerating over the past century will have a significant effect on global temperatures, precipitation, and sea level. Fossil fuel combustion by the developed countries over the past century is primarily responsible for the increase in global atmospheric concentrations of carbon dioxide. However, while energy use has peaked or declined in the developed countries, energy use is rising in China and other rapidly developing countries; if current trends continue, China will surpass the United States as the world's largest energy consumer and emitter of CO_2 early in the next century.

This report represents the summary findings and conclusions of a three-year study, *China: Issues and Options in Greenhouse Gas Emissions Control*. The study was conducted under a technical assistance project of the United Nations Development Programme, executed by the World Bank and implemented by the Chinese Government. As the first comprehensive country study funded by the Global Environment Facility (GEF), it addresses the key determinants of future energy use in China and identifies cost-effective options for limiting greenhouse gas emissions. The study was conducted by a joint team from the National Environmental Protection Agency of China, the State Planning Commission of China, UNDP, and the World Bank; the *Summary Report*, which is reproduced here, was subsequently endorsed by the respective institutions.

For the foreseeable future, China will need to expand the supply of electricity and other high-quality forms of energy to meet the needs of economic and social development. At the same time, there is vast potential for improving the efficiency of energy use in China through further institutional strengthening, the transfer of advanced technologies from abroad, and specific investments. An important way of improving energy efficiency is through the process of economic reform, including enterprise restructuring, financial sector development, and the move toward market pricing and efficient allocation of energy and other factors of production. Even with major improvements in energy efficiency, however, the only way to reduce greenhouse gas emissions in China over the longer term is through the expansion of renewable energy sources.

Because China and other large energy consumers figure so prominently in the calculus of global climate change, it is essential for them to take actions to limit greenhouse gas emissions, particularly those actions that have other positive economic or environmental benefits. Based on the recommendations of this study, the Chinese Government is proceeding with measures to reduce greenhouse gas emissions. For its part, and with the support of the GEF, the World Bank is committed to helping China improve the efficiency of energy use and to support the commercialization of cleaner alternative energy technologies.

Nicholas C. Hope
Director
China and Mongolia Department
East Asia and Pacific Region

Acknowledgments

This report summarizes the findings and recommendations of a United Nations Development Programme (UNDP) technical assistance study, "China: Issues and Options in Greenhouse Gas Emissions Control," supported by the Global Environment Facility and executed by the Industry and Energy Division, China and Mongolia Department, of the World Bank. It draws heavily on eleven subreports and numerous background papers prepared as inputs to the study by teams of Chinese and international experts.

In China, about twenty ministries and leading agencies worked on the study, and some 600 person-months were devoted to it. Research for the project in China was managed by the National Environmental Protection Agency (NEPA), which was the overall coordinator. The State Planning Commission (SPC) coordinated research on energy efficiency and alternative energy. Assistance was also provided by the Ministry of Machinery Industry and the Shanxi Planning Commission. International experts visited China on three major missions in April and May 1992, October and November 1992, and June and July 1993. A two-week workshop to design the China Greenhouse Gas Model was held at Stanford University in September 1992. Chinese experts visited the World Bank between October 1993 and March 1994, during which time the macroeconomic modeling was done and the major findings of the Summary Report were discussed and synthesized.

The Summary Report is the product of a joint Chinese-international study team including representatives from NEPA, the SPC, the UNDP, and the World Bank. Team members are listed on page iv. The joint study team attended a two-week workshop in Washington, D.C. in May 1994 to review the major findings and formulate conclusions.

The National Environmental Protection Agency and the State Planning Commission of China, the United Nations Development Programme, and the World Bank fully endorse the findings and recommendations of this report.

The Summary Report was drafted and edited by Todd M. Johnson (principal author), with Li Junfeng, Jiang Zhongxiao, and Robert P. Taylor (task manager). Editorial and production assistance was provided by Kathlin Smith.

Joint Chinese-international study team for the summary report

National Environmental Protection Agency of China

Zhang Kunmin, *Deputy Administrator*
Wang Hanchen (Project Coordinator), *Deputy Director, Modern Research Center for Economic and Environmental Policy*
Wu Baozhong, *Director, Department of Pollution Control*
Liu Yi, *Foreign Economic Cooperation Office*
Hao Jiming (Consultant), *Professor, Research Institute of Environmental Engineering, Tsinghua University*
Zhuang Dean (Consultant), *Senior Engineer, Chinese Academy of Science and Technology Development*
Chen Xikang (Consultant), *Professor, Institute of Systems Science, Chinese Academy of Science*
Xu Deying (Consultant), *Professor, Research Institute of Forestry, Chinese Academy of Forestry*

State Planning Commission of China

Shen Longhai, *Director, Department of Spatial Planning and Regional Economy*
Zhou Fengqi, *Director, Energy Research Institute*
Zhu Liangdong, *Advisor and Senior Engineer*
Zhou Changyi, *Division Chief, Department of Raw Materials Industry*

Wang Shumao, *Division Chief, Energy Conservation and Rational Utilization Division, Energy Research Institute*

Liu Zhiping, *Associate Professor, Energy Research Institute*

Li Jingjing, *Engineer, Energy Research Institute*

United Nations Development Programme

Susan McDade (UNDP Project Manager), *Assistant Resident Representative (Beijing)*

Richard Hosier, *Principal Technical Advisor on Climate Change, Global Environment Facility*

The World Bank

Robert P. Taylor (Bank Task Manager), *Senior Energy Economist, EA2IE*

Todd M. Johnson (Study Principal Investigator), *Environmental Economist, EA2IE*

Li Junfeng (Consultant), *Energy Efficiency Specialist, EA2IE*

Jiang Zhongxiao (Consultant), *Macroeconomist, EA2IE*

Robert M. Wirtshafter (Consultant), *Energy Efficiency Specialist*

Abstract

This report summarizes the findings and recommendations of a three-year study on greenhouse gas (GHG) emissions and options for abatement in China over the coming decades. Macroeconomic modelling results show that the continuation of rapid economic growth in China could result in a threefold increase in GHG emissions between 1990 and the year 2020. Specific measures for limiting GHG emissions are examined in detail, including improvements in energy efficiency, more rapid introduction of non-fossil energy technologies, afforestation for carbon sequestration, and modifications to various GHG-producing agricultural practices. In the short- to medium-term (before 2010), energy efficiency holds the greatest potential for low-cost GHG emission reduction. Over the longer term, however, the only option for China and the world is to switch to non-carbon energy sources. The study concludes that a two-pronged strategy for reducing GHG emissions in China should be adopted, whereby (i) economic reform and policy initiatives are continued for the purpose of improving resource allocation and encouraging energy conservation, and (ii) a set of priority investment and technical assistance programs are undertaken which promote the acceleration of more efficient and low-carbon technologies and which improve the institutional and human resource capacities to implement and sustain these programs.

Data Notes

Currency equivalents

All Chinese currency (yuan) references in this report are in constant 1990 prices and are converted to U.S. dollars at the official exchange rate of 4.7 yuan (Y) per U.S. dollar.

Weights and measures

1 ton of coal = 0.7143 tce (average) = 20.934 GJ
1 ton of crude oil = 1.43 tce = 41.816 GJ
1000 m^3 of natural gas = 1.33 tce = 38.931 GJ
1 ton fuelwood (air dry) = 0.54 tce
1 hectare = $10^4 m^2$ = 2.47 acres
1 GJ coal = 25.8 kgC
1 GJ crude oil = 20 kgC
1 GJ natural gas = 15.3 kgC
All CO_2 weights expressed as molecular weight of carbon (C)

Abbreviations and acronyms

C	carbon
CFC	chlorofluorocarbons
CH_4	methane
CO	carbon monoxide
CO_2	carbon dioxide
EIRR	economic internal rate of return
FGHY	fast-growing high-yield (forestry plantations)
FIRR	financial internal rate of return
GDP	gross domestic product
GEF	Global Environment Facility
GJ	gigajoule
GHG	greenhouse gas
GWP	global warming potential
I-O	input-output
IPCC	Intergovernmental Panel on Climate Change
kgce	kilogram of coal equivalent
kW	kilowatt
kWh	kilowatt-hour
LNG	liquified natural gas
LPG	liquified petroleum gas
mtce	million tons of coal equivalent
MW	megawatt
N_2O	nitrous oxide
NEPA	National Environmental Protection Agency of China
NOx	nitrogen oxides
NPV	net present value
OECD	Organization for Economic Cooperation and Development
PV	photovoltaic
SO_2	sulfur dioxide
SOE	state-owned enterprise
SPC	State Planning Commission of China
t	metric ton
tce	ton of coal equivalent
TSP	total suspended particulates
TVE	township and village enterprise
UNDP	United Nations Development Programme

Executive Summary

INTRODUCTION

The burning of fossil fuels and other human activities are changing the balance of CO_2 and other heat-trapping gases in the atmosphere. According to scientific theory, the "greenhouse effect" has the potential to dramatically alter the earth's climate in a relatively short span of time. At current emission rates, global atmospheric CO_2 concentrations will double by the middle of the twenty-first century. According to the Intergovernmental Panel on Climate Change (IPCC), this will result in a warming of the earth's atmosphere by 1.5–4.5 degrees C and cause global mean sea levels to rise by 0.25–0.50 meters.

One of the goals of the Framework Convention on Climate Change (FCCC), established at the Environmental Summit in Brazil in 1992, is to provide assistance to developing countries to help them limit their emissions of greenhouse gases (GHGs). Despite the need to take measures to limit GHG emissions, economic development and poverty alleviation are recognized by the FCCC as the top priorities for China and other low-income countries.

Energy is the largest source of greenhouse gas emissions worldwide, and China currently accounts for 10 percent of global energy use. However, among countries with the highest GHG emissions, only China is likely to maintain rapid rates of economic growth well into the next century and this expansion will require additional energy. This study evaluates the reduction potential and costs of a broad array of options for reducing GHG emissions in China over the next twenty-five years. This summary report is the result of a two-year investigation by a joint Chinese-international study team of representatives from the National Environmental Protection Agency and the State Planning Commission of China, the United Nations Development Programme, and the World Bank, with support from the Global Environment Facility (GEF).

Potential climate change impacts on China

China and other low-income countries are likely to be most affected by climate change because their economies are more dependent on climate-sensitive sectors, such as agriculture, and because they are least able to afford mitigation or adaptive measures, such as the building of dikes. Chinese researchers have predicted that a doubling of CO_2 will have a negative impact on rice, wheat, and cotton production because of the combined effects of higher temperatures, increased soil evaporation, and more frequent and severe storms. This prediction is consistent with recent modeling work by the IPCC, which estimates that agricultural production would fall by 6–8 percent worldwide, and by 10–12 percent in developing countries, with a doubling of atmospheric CO_2 concentrations. Rising sea level is another concern for China and other countries with large populations living in low-lying coastal plains. According to studies by Chinese researchers, a 1-meter increase in sea level, when combined with storm surge and the astronomical tide, will flood areas below a 4-meter contour line in China's coastal plains. The inundated land would cover an area the size of Portugal, including the cities of Shanghai and Canton, and would displace 67 million people at current population levels, more than the population of the Philippines.

Despite the potential for disastrous consequences associated with human-induced climate change, there is substantial uncertainty in impact predictions and in the climate change process itself. Until the scientific uncertainty associated with the greenhouse effect and its impacts can be reduced, it is rational for China to focus first on a "no-regrets" policy for GHG reduction; that is, those policies or projects that make sense for economic, social, or environmental reasons other than GHG reduction. A no-regrets policy is particularly important for low-income countries, where there are many other urgent developmental needs and environmental concerns.

Estimates of current GHG emissions

Energy consumption is by far the largest contributor of GHG emissions in China, accounting for more than 80 percent of total emissions in 1990.[1] Coal accounted for 76 percent of primary energy use in China in 1990, followed by oil (17 percent), hydropower (5 percent), and natural gas (2 percent). Industry and electric power generation accounted for 75 percent of China's CO_2 emissions from commercial energy consumption, while the residential sector accounted for 14 percent and transportation 4 percent. Industrial boilers used outside the power sector consumed more than 350 million tons of coal in China in 1990, accounting for about 35 percent of the country's coal use.

Methane accounted for approximately 13 percent of China's GHG emissions in 1990 (CO_2 equivalent), with rice fields, coal mining, ruminant animals, and animal wastes contributing 88 percent of methane emissions (Figure 1). All other sources of GHG emissions, including CO_2 from cement manufacturing, methane from landfills, N_2O from fertilizer, and non-CO_2 emissions from biomass burning and land use changes, accounted for about 6 percent of China's total GHG emissions in 1990. Forests were a small sink of carbon in 1990.

METHODOLOGY

The study entailed extensive macroeconomic and microeconomic analysis, which was conducted by international and Chinese teams. The macroeconomic work focused on (i) detailed sector and subsector analysis of energy use and GHG emission trends, and the potential for energy conservation, interfuel substitution, and GHG control in the forestry and agriculture sectors; (ii) development of scenarios for future growth in GHG emissions; and (iii) identification and analysis of the factors that have the greatest

---

[1] The greenhouse gases that have been estimated in this report—CO_2, CH_4, and N_2O—have been added based on their heat-trapping properties, or "global warming potential" (GWP). For an explanation of the use of GWPs, see *Climate Change, The IPCC Scientific Assessment*, Intergovernmental Panel on Climate Change, (Cambridge University Press, 1990).

Figure 1 China: Greenhouse gas emissions, 1990

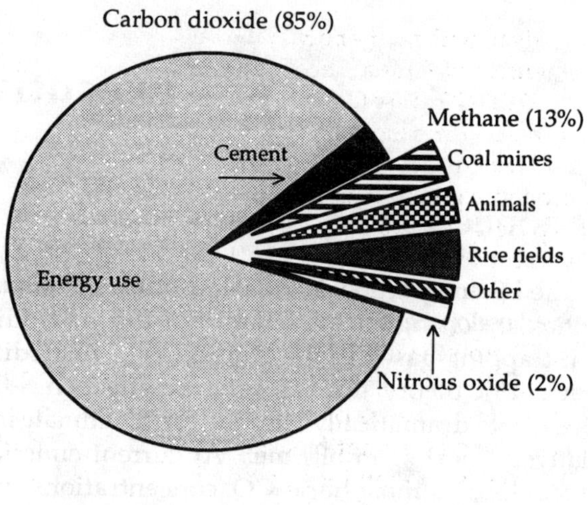

Note: Excludes CFCs. In 1990, forests were a small sink of carbon dioxide.

influence on future emissions levels. A China GHG model was constructed for the study, composed of an econometric macroeconomic model, an 18-sector input-output model of the economy, a matrix of energy demand coefficients, and an emissions matrix to calculate both global (CO_2, CH_4, and N_2O) and local (TSP and SO_2) emissions. Economic structure and its change over time is represented by a series of input-output tables projected to the years 2000, 2010, and 2020, based on China's historical situation and the trends in other countries, including Japan, the United States, and Germany.

In the microeconomic work, cost-benefit analysis was used to assess the net unit cost of GHG reduction for several reduction options: benefits were subtracted from the costs of various projects, with the remaining "net" cost divided by the discounted tons of carbon equivalent reduced. Financial, economic, and, where possible, local environmental economic analyses were conducted for twenty-five energy efficiency projects, alternative energy sources, timber and protective afforestation plantations, and selected agricultural programs. The net present value, internal rate of return, and cost per ton of carbon equivalent reduced were calculated for these projects.

Some of the conclusions of this study may need to be revised as more information about climate change, its effects, mitigation options and technologies, and China's economic development path becomes available.

FACTORS AFFECTING GHG EMISSIONS AND OPTIONS FOR REDUCTION

Even with further dramatic declines in the energy intensity of China's economy, continued economic growth will cause China's greenhouse gas emissions to increase substantially by 2020. Under a *high-growth scenario* developed by the study team, China's economy grows at an annual rate of 9.5 percent during the 1990s, 8.0 percent from 2000 to 2010, and 6.5 percent from 2010 to 2020. In the team's Baseline GHG Scenario, modeled to depict a continuation of the development trends of the 1980s and early 1990s, rapid growth would be accompanied by a three-fold increase in GHG emissions between 1990 and 2020. The primary cause of the GHG increase would be a rise in coal consumption from 1.05 billion tons in 1990 to about 3.1 billion tons in 2020. Actual levels of future coal use and GHG emissions can be influenced by a variety of policies and investment patterns. The key factors are summarized in the following section and described in greater detail in Chapter 2.

Macroeconomic factors

Rate of economic growth. Slower growth in GDP can be expected to bring slower growth in GHG emissions. However, the rate of economic growth is also correlated with improvements in energy efficiency. More rapid development increases the opportunity for adopting new, more energy-efficient processes and technologies. In the team's *slower-growth scenario*, average annual GDP grows at 8.5 percent during the 1990s, 6.5 percent from 2000 to 2010, and 5.0 percent from 2010 to 2020. While China's total GDP in 2020 under the slower-growth scenario is less than 70 percent of the GDP in the high-growth scenario, GHG emissions decline by only 10 percent.

Economic structure. Changes in the structure of China's economy will continue to be primarily responsible for future declines in energy intensity. In terms of energy use per unit GDP, China is one of the most energy-intensive economies in the world. However, the country's energy intensity is expected to drop sharply because of changes in the product mix and sources of value added in industry. With further economic reform, the increase in industrial value added, which is expected to drive China's growth, will come less from quantitative increases in the output of basic industrial goods and more from greater product diversification, specialization, and improvements in quality.

The Baseline GHG Scenario results in an energy use/GDP growth elasticity of about 0.5 from 1990 to 2020, and a reduction in the energy intensity of GDP in 2020 to one-third the 1990 level. The macroeconomic analysis found that structural factors, or "indirect energy savings," account for about three-quarters of the decline in total energy intensity during the period.

Specific reduction options

Energy efficiency. In addition to indirect energy savings through structural change, there is great potential for direct energy savings by reducing energy use per unit of physical output. This can be accomplished by the following means:

- *industrial modernization and restructuring*, including adoption of more efficient industrial processes, achievement of economies of scale, and improved management. In particular, an increase in the scale of plants in the rapidly expanding thermal power sector can yield major efficiency gains.

- *"classic" industrial energy conservation projects*, including improved waste heat, gas, and waste stream recovery; expanded use of cogeneration; industrial furnace and kiln renovation; installation of improved monitoring and control systems; and use of improved insulation and other renovations in thermal and steam systems.

- *improvements in the energy efficiency of new models of widely used equipment*, particularly small and medium-sized coal-fired industrial boilers, electric motors and associated industrial electrical equipment, air conditioning equipment and refrigerators, lighting devices, and steam traps associated with industrial piping networks.

- *additional coal processing and improvement of the quality and consistency of coal supply*, including expansion of coal gasification, washing, screening and sorting, and briquetting.

- *residential and commercial sector energy-efficiency measures*, including the use of improved building designs, building materials, centralized heating systems, residential stoves, and electrical equipment.

Alternative energy. In addition to improved energy efficiency, the other major option for large-scale reduction of GHGs in China—and the world—is the substitution of low- or non-carbon alternatives for fossil fuels. Reliance solely on market forces is not likely to result in the substitution of large amounts of low-carbon fuels for coal in China over the medium term. Accelerated development of alternative energy sources, in line with their strategic importance for fuel diversification, as well as for improvements in environmental quality and poverty alleviation, will require a program of well-targeted government and international support for technology development, especially for renewable energy. By the year 2020, large-scale adoption of low-carbon energy sources such as hydro, nuclear, gas, biomass, solar, and wind technologies could displace a substantial quantity of coal, particularly for electric power generation. Based on a projection of current trends in technological development, the study team estimates that low-carbon fuels could provide 35–40 percent of electric power generation and 15–20 percent of China's energy supply by the year 2020. Displacement of coal in non-power uses, such as industrial process heat and residential cooking, will be a particular challenge.

Forestry. By instituting a massive afforestation program, China could reduce net GHG emissions by about 10 percent in 2020 compared with the Baseline GHG Scenario. To achieve this level of carbon sequestration, China would have to increase forested land by 4–5 million hectares per year between now and the year 2020, extend the use of fast-growing, high-yield plantations, and broadly disseminate advanced silviculture techniques. This level of planting would increase the percentage of forested land in China from about 13 percent in 1990 to more than 20 percent by the year 2020. Although fuelwood plantations do not sequester much carbon on a net basis, they can contribute to GHG reduction by producing biomass, a substitute for fossil fuels. Utilizing wood and forest residues grown on a sustainable basis to generate power is an effective way to reduce GHGs by both replacing coal and increasing carbon sequestered on forest stands.

Agriculture. By further disseminating techniques for cattle breeding and raising that improve the efficiency of food digestion and use, methane emissions from animals in China could be reduced by 25 percent by the year 2020. Rice growing techniques that decrease the amount of time rice fields are flooded could reduce methane emissions from rice cultivation in China by 15–20 percent by the year 2020.

Cost issues aside, the study team estimates that the increase in GHG emissions in China accompanying the high-growth GDP scenario for 1990–2020

Table 1 Potential for GHG reduction compared to baseline scenario, 2020

	mtC	Share %
Energy efficiency (high scenario)	-330	41
Alternative energy (high scenario)	-237	29
Afforestation (high scenario)	-221	27
Agricultural programs	-15 to -25	03

could be limited to a doubling, rather than a tripling, compared with the Baseline GHG Scenario, which estimates 2,398 mtC. This could be achieved through the adoption of more aggressive direct energy efficiency measures, expanded use of alternative energy, massive afforestation, and changes in selected agricultural techniques.

THE COSTS OF GHG EMISSION REDUCTION

The most cost-effective options for reducing GHG emissions are those in which other, non-GHG-related project benefits exceed project costs. China is expected to have ample opportunity for implementing such "no-regrets" options into the next century. A key challenge is to overcome barriers that impede rapid implementation of these projects.

Energy efficiency. The study team evaluated twenty-five representative energy-efficiency projects in China, primarily in the industrial sector, that show promise for reducing energy use and GHG emissions. Overall, the financial and economic internal rates of return of these investments were attractively high, even without consideration of their major GHG reduction benefits. Most projects also had human health benefits through reductions in TSP and SO_2 emissions. Yet, progress in implementing many of these projects is slow. Industrial modernization projects require large amounts of upfront investment capital and often carry substantial risk. In today's growing economy, enterprise managers typically pay little attention to the life-cycle returns of cost-saving investments and tend to be interested in smaller energy-efficiency projects only if payback periods are short. Although important in the aggregate, the cost savings of some energy conservation measures are too small to be important to individual enterprises. Finally, access to advanced foreign technologies and knowledge of the opportunities and experiences of other enterprises at home and abroad is often poor.

Alternative energy. Most of the low-carbon fuels that could substitute on a large scale for coal are judged to be more costly than coal over the next twenty-five years, even when the costs of meeting strict environmental standards for coal use are added. There is some opportunity for expanding "no-regrets" investments in current non-coal energy supply technologies, including investments in hydroelectric power, the use of coal-bed methane, fuelwood production under favorable natural conditions, expanded exploration and development of natural gas, and wind-powered electricity generation. Based on the cost expectations of most experts today, however, expanded development of nuclear power and other non-coal alternatives would be extremely expensive. To reduce the share of coal in the Baseline GHG Scenario from 67 percent in 2020 to 57 percent would carry an additional cost of more than US$100 billion.

Forestry. A net cost analysis of carbon sequestration from forestry projects found that four types of plantations in China are financially and economically attractive on a life-cycle basis even if GHG benefits are not considered: (i) intensively-managed, fast-growing, high-yield (FGHY) timber plantations on good land; (ii) extensively-managed timber plantations in South and Southwest China; (iii) improved open forest management regimes in South China; and (iv) intensively-managed FGHY fuelwood plantations in South and Southwest China.

Agriculture. Certain improved breeding and feed programs were found to have very high financial and economic rates of return and are being adopted across China. An ammoniated feed program is being promoted because it results in less local air pollution from crop residue burning as well as yielding high financial returns. Economic analyses were not performed on the two rice-cultivation practices, but these techniques are already being used in some parts of China for reasons other than GHG reduction.

RECOMMENDATIONS

Based on its extensive analysis, the joint study team concludes that the following elements are most important in forming an optimal strategy to reduce GHG emissions in China:

1. continuation and expansion of the economic reform program to improve the overall efficiency of resource use;

2. accelerated implementation of "no-regrets" projects over the short to medium term; and

3. expanded development of less-carbon-intensive energy technologies for the longer term.

Economic reform

Completion of a successful transition to a market economy is fundamental to the Chinese government's economic development policy. Success in economic reform is also a key factor influencing future GHG emission levels. The pace of structural change in the economy will be largely determined by progress in economic system reform, and the pace of structural change is the most important factor influencing the energy efficiency of the economy. In addition, continued progress in economic reform is important to help spur technical energy conservation and other no-regrets projects. Further development of enterprise autonomy and accountability, hard enterprise budget constraints, expanded competition, and completion of price reforms are all important elements of a framework of incentives for improving energy efficiency.

As in developed countries, environmental regulatory policy can be an important tool in China to encourage enterprises to use clean forms of energy and to use energy more efficiently. While China has made significant progress in establishing a comprehensive environmental regulatory system, enforcement must be improved and modifications are needed to make it more effective in a market system. Internationally, China should adopt and disseminate its strategy for GHG reduction as part of its participation in global climate change initiatives.

Priority policy, investment, and technical assistance areas

While continued economic reform is critical for effective GHG abatement, China's GHG control strategy must also include a series of specific actions. These include reforming certain sector-specific policies; investing in no-regrets projects; and improving institutional, technical, and managerial capacities.

The joint study team concludes that the project areas summarized in the following paragraphs and in Chapter 4 represent areas of high priority for action to reduce GHG emissions given the priority that China correctly places on economic development.

Primary emphasis should be placed on measures to speed investment in projects that have substantial benefits aside from GHG reduction. The bulk of investment in these projects should come from enterprises themselves. The challenge for the Chinese government is to assemble effective packages of policy reform, investment, and institutional strengthening measures to reduce barriers to implementing these projects and to catalyze a broad-based response. Utilizing international assistance where necessary, measures to speed implementation of no-regrets projects or to invest in the development of more cost-effective low-carbon energy supply technology in China are among the most cost-effective means available globally to reduce GHG emissions.

Improvements in energy efficiency. Improvement in the technical efficiency of energy use is the highest priority for action to mitigate GHG emissions in China over the short and medium term. In terms of policy, further enterprise reform and improved local environmental regulation are important. Further efforts also are needed to complete energy price reform, such as the rationalization of natural gas prices. After reviewing the impact of the recent liberalization of coal prices, the Chinese government should weigh the potential advantages and disadvantages of coal taxation, especially as a tool for meeting local environmental objectives. Finally, there is a need to promote the implementation of more effective energy-efficiency standards for equipment, especially boilers and key electrical equipment.

In addition, the Chinese government should expand its efforts to remove barriers to "no-regrets" projects, such as insufficient access to information about technical opportunities and experiences, lack of access to foreign technology, abnormally high technical and market risks associated with new processes or technologies, high transaction costs for small investments, and institutional constraints. Recommended means to overcome these barriers, particularly in the current period of economic transition, include the following: (i) improved credit facilities for energy conservation investments, emphasizing financially viable investments with longer payback periods; (ii) well-targeted concessional finance for demonstration of new energy-saving technologies, including those from abroad that carry substantial technical or market risk; (iii) development of energy service companies which bear the risk of energy-saving investment in enterprises in return for a share of the financial return; (iv) better dissemination of information on energy conservation investments, em-

phasizing financial benefits and targeting small enterprises, including TVEs; and (v) technical assistance and training, including energy auditing, preinvestment analysis, and staff training.

Strategically important areas of concentration include the following:

• industrial energy conservation projects, especially those focusing on more efficient heat, steam, and by-product gas utilization.

• projects to develop, manufacture, and effectively market more efficient models of energy-consuming equipment. National policies should be implemented to encourage the development of higher-efficiency small- and medium-scale coal-fired industrial boilers.

• improvement in the quality of coal supply, over the near term, through increased washing of steam coal, better sorting and matching of coal types and sizes, and briquetting and pelletization. This requires policy and investment support, beginning with a pilot project in one or two localities. Adoption of improved coal gasification technology will be important over the long term.

• improvement in the energy efficiency of residential and commercial buildings. This can be achieved through the implementation of policies to overcome institutional barriers; the promotion of more efficient use of energy-intensive materials such as cement, steel, and bricks; the development, demonstration, and marketing of new energy-saving construction products; and the integration of more efficient district heating system designs in housing projects.

Although the transport sector is projected to account for only 5 percent of CO_2 emissions from China in 2020 under the baseline scenario, the absolute amount of energy consumed by the sector would increase from 45 mtce in 1990 to 173 mtce in 2020 and options for reducing emissions through efficiency improvements, modal shifts, and structural change could be important globally. Further research on such options is needed.

Alternative energy development. Greater support for the development of low- or non-carbon energy technologies is urgently needed now for non-coal energy alternatives to play a major role in China's economy over the medium and long term. China will require massive quantities of alternative energy supply during the early part of the next century for local environmental and logistical reasons, as well as for reasons relating to the global environment. Given the abundance of low-cost coal resources in China, the necessary technological development of alternative energy sources is not likely to occur at sufficient speed solely through reliance on market forces.

Accordingly, the joint study team recommends that the government establish, with international assistance where required, an aggressive program to accelerate the technological development of alternative energy sources, particularly renewable energy technologies. Primary emphasis should be placed on technologies that have the potential to make a large contribution to China's long-term energy supply. The program should focus on research, technology transfer from abroad, and technology demonstration and dissemination activities aimed at reducing the costs of the alternative energy supply and improving its cost-effectiveness when compared with the use of coal.

In addition, support should be given to speed the adoption of alternative technologies where applications are currently cost-competitive with coal; for example, in certain coal-bed methane, biomass, wind, or solar energy applications. Expanded exploration of natural gas should also be emphasized.

GHG control in the forestry sector. Afforestation and forestry management practices that have the potential for maximizing carbon sequestration at the lowest net cost should be the focus of government support. Policies to encourage private investment in the forestry sector are especially important, including improvements in rural capital markets, further price reform, clarification of property rights, and liberalization of foreign trade and investment policies. Technical assistance or technology transfer can also help expand China's fast-growing high-yield plantation program and improve silviculture techniques, nursery management, and forestry research and extension.

GHG control in the agricultural sector. In the agricultural sector, China should focus its GHG control efforts on expanding and accelerating programs that increase the efficiency of livestock production. Meanwhile, the international community should support research in China on emissions and control mechanisms for methane from rice fields and N_2O from fertilizer, and the development and dissemination of related applicable no-regrets technologies. Govern-

ment efforts should focus on policies to improve rural credit and expand technical assistance, demonstration, and dissemination of proven no-regrets programs.

International assistance for reducing GHG emissions in China

Development assistance. Conventional international and bilateral assistance to China will continue to be important for improving resource allocation in general and energy efficiency in particular, both of which are essential for GHG reduction. The technical assistance and lending programs of international agencies have helped to advance energy price reform, capital market development, enterprise management, and ownership reforms in China. Continued and expanded support from international agencies is also needed for China to implement many of the priority investments in GHG reduction outlined above, such as energy conservation, efficient power development, high-yield timber plantations, improved animal feed, and alternative energy technologies. In addition, because private sector foreign investment will be important in the modernization of China's capital stock over the next several decades, the Chinese government should seek to maximize the efficiency of foreign technologies and processes that are brought to China.

Role of the Global Environment Facility. GEF resources should be used to promote and accelerate priority policies, investment, and technical assistance projects for GHG reduction in China outlined above and in Chapter 4. Although the exact criteria for future GEF projects have yet to be finalized, based on experience with the GEF to date, the joint study team expects that the following principles will guide the selection of future global climate change projects: (i) GEF resources should be used to advance global environmental objectives and not as another source of development funding; (ii) projects seeking GEF support should be of national priority and part of the country's overall climate change strategy; and (iii) GEF resources should aim to maximize the amount of GHG reductions per unit of GEF funding.

To support China's priority policies, investment, and technical assistance projects for GHG reduction, GEF resources should be used primarily to overcome market and non-market barriers to implementing "no-regrets" projects and for accelerating the development of promising alternative energy technologies for the longer term. Whether the constraints are of a technical, informational, or institutional nature, the use of GEF resources to overcome barriers to no-regrets projects will result in exceptionally large reductions in GHG emissions per unit of GEF funding. The joint study team believes that the use of GEF resources in this manner will provide the most benefits to China and be the most efficient use of GEF funds for GHG reduction.

Where technical or market risk represents a current binding constraint to the adoption of high-efficiency technologies, the GEF can play a critical role in supporting technology transfer through the purchase of technology rights, joint Chinese-international pilot project development programs, and the implementation of demonstration projects. The GEF should also support projects that address institutional, informational, or policy constraints, such as the development and implementation of more effective standards, dissemination of proven processes and techniques, training programs, and public education campaigns. In addition to project options that have potential for broad dissemination under current market conditions but which require strategic input from GEF to overcome existing constraints, GEF resources should also be used to accelerate alternative energy technology development, particularly for renewables, so that such technologies can be adopted on a large scale in the future at costs competitive with carbon-intensive fuels.

As an initial assessment of GHG mitigation for China, this report provides a framework for identifying priority projects for future GEF eligibility and funding in China.

Chapter 1

Study Objectives and Background

STUDY OBJECTIVES

Scientific research and the historical climate record appear to support the theory that the earth's climate is being affected by anthropogenic emissions of greenhouse gases. Given the potential risks of global climate change, it is appropriate for countries to limit emissions, or as one author has stated, to "buy greenhouse insurance."[1] However, because GHG emissions are usually closely correlated with economic growth, limiting emissions is tantamount to limiting economic development—an unwelcome prospect for low-income countries. Because there are other important developmental objectives that need to be funded, nations should buy the least expensive insurance available, particularly until the significant uncertainty associated with global climate change is reduced.

This report identifies and evaluates low-cost options for GHG reduction that China can implement in the short- to medium-term—between now and the year 2010. Of the major contributors to worldwide GHG emissions, only China is expected to experience rapid economic growth over the next several decades. Given China's size and developmental potential, major reductions in GHG emissions in the rest of the world will be required to offset even modest increases in emissions from China. By the same token, modest improvements made now to reduce emissions from China will have a tremendous impact on worldwide GHG emissions over time.

A major objective of the Climate Change Convention, which was adopted at the 1992 Environmental Summit in Brazil, is to provide a mechanism for the transfer of financial resources and advanced technologies to developing countries to assist them in reducing GHG emissions. To ensure that these resources are used efficiently, it is essential that countries prepare climate change strategies that identify the least-cost options for reducing GHG emissions. This report, which has been prepared jointly by the World Bank and the Chinese government, is intended to identify priorities for GHG reduction that could be part of China's overall climate change strategy.

The tasks for the study were as follows:

• Establish a baseline GHG emissions inventory for China for 1990 based on internationally agreed-upon standards.

• Estimate GHG emissions over the next three decades (1990–2020) under different scenarios.

• Identify the factors that will have the greatest impact on reducing GHG emissions and quantify the magnitude of potential reductions.

• Develop a methodology for calculating the net cost of GHG emissions reduction so that different options can be compared.

• Identify and evaluate emissions reduction options in four subsectors: energy-use efficiency, alternative energy, forestry, and agriculture.

• Identify barriers to the development of least-cost options for GHG reduction and describe what steps can be taken by China, international development organizations, and the Global Environment Facility (GEF) to implement a least-cost reduction strategy in China.

---

[1] Alan S. Manne and Richard C. Richels. *Buying Greenhouse Insurance: The Economic Costs of CO_2 Emission Limits* (Cambridge: MIT Press, 1992).

POTENTIAL CLIMATE CHANGE IMPACTS IN CHINA

Concern over the greenhouse effect is based on the potential damage that global climate change could inflict on food production, human settlement, and terrestrial ecosystems that support both plant and animal life. There is considerable uncertainty in estimating the impacts of anthropogenic contributions of CO_2 and other heat-trapping gases in the atmosphere. One of the tools used to explore future climate change is the computerized General Circulation Model, which is used to simulate past and future climate variation. Such models have been used to simulate the effect of increasing concentrations of CO_2 in the atmosphere. According to the leading models, a doubling of atmospheric CO_2, which at current levels of emissions would be reached before 2050, would increase global mean surface temperatures by 1.5–4.5 degrees centigrade, while the global mean sea level would rise by 25–50 cm.[2]

While the impacts of future climate change are hard to predict, several studies have been concluded or are underway that explore potential global impacts. Most at risk from climate change will be communities that are least able to adapt and sectors that are most dependent on climate, such as agriculture. According to simulations done by the Intergovernmental Panel on Climate Change (IPCC), a doubling of CO_2 will lead to a decline in global agricultural production of 6–8 percent, while in the developing countries the decline would be on the order of 10–12 percent. One of the most serious impacts will be the change in water availability. For instance, with a 1–2 degree centigrade temperature increase and a 10 percent decrease in precipitation, water runoff could decrease by 40–70 percent per year. Because there are large populations in Asia living in low-lying coastal areas, there is particular concern in the region over sea-level rise and the related issues of land subsidence, salt water inundation of fresh waters, and the increased frequency and severity of storms.

Although predictions of *regional* climate change effects are less certain than global mean levels, estimates of changes for China with a doubling of atmospheric CO_2 have been simulated through the use of a leading climate change model (Table 1.1). Using these results as a guide, and based on historical climate change data for different parts of China, Chi-

nese researchers have predicted four principal impacts for China from climate change:[3] i) a general warming trend which would extend China's tropical and northern growing regions, ii) lower crop yields due to reduced water availability caused by increased evaporation, iii) a greater threat of soil erosion caused by higher precipitation levels and a decline in soil moisture, and iv) increased flooding of coastal and low-lying plains caused by sea-level rise and more frequent and severe storms.

Table 1.1 Simulated changes in China's climate with a doubling of CO_2

	Winter	Summer
Surface air temperature (degrees centigrade)	4.8	3.9
Precipitation (%)	12.7	9.3
Soil moisture (%)	-2.4	-2.9
Total cloud cover (%)	-5.1	-2.9

Source: Wang et al., 1992, National Center for Atmospheric Research, Albany, NY.

The wide range of climatic zones in China— subfrigid in the north to tropical in the south; humid in the east to arid and alpine in the west—would be affected in different ways by global climate change. The warming would be greater in winter than in summer. Rainfall would increase in the warming areas of the north and west as well as in coastal areas. Summer flooding would become more frequent in the areas of the Yangtze, Liaohe, and Huaihe Rivers, as would typhoons and storms, which affect China's southeast coast. Many parts of China could experience an increase in drought, hot dry winds, and soil evaporation, especially in the spring and early summer.

Agricultural production

In China, agriculture is the economic sector that would be most affected by climate change. Grain and vegetable crops, livestock, fisheries, and forest products would be affected by global warming in both

---

[2] *Climate Change: The IPCC Scientific Assessment*, IPCC, (Cambridge: Cambridge University Press, 1990).

[3] For details, see the project subreport prepared for this study, *Potential Impacts of Climate Change on China*, September 1994.

positive and negative ways. With a doubling of CO_2, the temperature zones in China are predicted to move north by as much as 4 degrees latitude. This would reduce frost damage to crops, and, assuming that all other factors remained constant, would allow an increase in multiple cropping patterns and aid in photosynthesis through the CO_2 fertilization effect.[4] As temperatures rise, however, soil evaporation increases; studies in China have shown that for every 1 degree centigrade increase in temperature, soil evaporation increases by 5–10 percent. As a result, many parts of China would become drier with a doubling of CO_2. In regions where evaporation is greater than precipitation, the need for additional irrigation would increase the threat of secondary soil salinization. Most areas of China would be more vulnerable to soil erosion as a result of climate warming. Sea-level rise would inundate some coastal areas; low-lying fields would be flooded more often, storm surge would be more frequent, typhoons would be more severe, and drainage of low-lying areas would be more difficult. In delta areas, sea water intrusion would degrade farmland and cause the salinization of groundwater. As temperatures increase, insects, rodents, and weeds could appear earlier in the spring and remain longer in the autumn, causing more damage to crops and increasing pest and weed control costs.

Grain and vegetable crop production. Although global warming would probably increase the area available for triple-cropping of rice, the average yield is projected to fall because of decreased water availability. At the margins of cereal production in the north, increased temperatures would increase crop yields, particularly in Northeast China and on the Qinghai-Tibet Plateau. Based on the many factors affecting crop production, including temperature, precipitation, CO_2 fertilization, evaporation, storms, flooding, disease, pests, and crop adaptability, Chinese researchers have estimated a decline in productivity for several crops, including wheat, rice, and cotton.

Livestock. The production of grasses, the primary feedstuff for animals in China, is predicted to be adversely affected by climate change. While there would probably be marginal increases in grass pro-

duction in parts of North China, overall grass production in China is predicted to decline by the year 2030 because water will be more scarce. The result would be a decline in the number of animals that could be supported on the same area of grassland, or an increase in costs for supplementing animal diets.

Fisheries. Several climate change variables were examined for their impact on both inland and coastal fisheries in China, including changes in temperature, water area, availability of food nutrients, and storms. While some fisheries would benefit from increases in temperature and rainfall, the overall impact is predicted to be a decline in production by the year 2030. Particularly affected would be the Yangtze River region, which accounts for 50 percent of China's inland water area. There, lower temperatures in winter and higher frequency of storms and flooding could decrease fishbreeding considerably by the year 2030.

Sea-level rise

The most serious impact of a rising sea level would be the increased frequency and severity of storm surges and typhoons. However, sea water intrusion and coastal erosion would also cause economic and social losses for China. In addition to the 25–50 cm rise in global sea level by 2050 caused by global warming, relative sea level in parts of China is predicted to rise because of tectonic subsidence and groundwater overpumping. By 2050, relative sea level is predicted to rise 70–100 cm in Tianjin, 50–70 cm in Shanghai, and 40–60 cm in the Pearl River Delta. Chinese researchers have shown that if coastal areas are not protected, a 1-meter increase in sea level will flood regions lying below a 4-meter contour line in China's coastal plains when combined with the astronomical tide and storm surge. The inundated area would cover fourteen cities and counties in the Pearl River Delta, including Canton, and thirty-four cities and counties in East China, including Shanghai. This area covers a total of 92,000 km² and the flooding would displace 67 million people at current population levels.

Terrestrial ecosystems

Wetlands, forests, and deserts, and their respective flora and fauna, would most likely be adversely affected by a doubling of CO_2. China's wetlands, 40 percent of which already face medium to serious threat according to the Asian Wetlands Bureau, would be further affected by increased evaporation and nitrification resulting from the temperature rise. More

---

[4] While laboratory experiments show that increased CO_2 concentrations can increase the productivity of certain types of plants by as much as 50 percent, this initial response declines over time and productivity eventually approaches that of crops grown under normal CO_2 levels.

than 500 species of freshwater fish and 300 species of birds inhabiting Chinese wetlands and shallow inland waters would be affected by changes in the area, seasonality, and location of wetlands. Mangroves, an important component of coastal ecosystems, could move north with warmer temperatures; however, they would also be threatened by sea level rise and more frequent tropical storms. Arid and semi-arid regions will be most affected by climate change, experiencing decreases in rainfall and soil moisture and increases in temperature. Combined with the harmful effects of doubled CO_2 on plant growth in these areas, desertification could increase. The area suitable for growth of tropical rain forests would increase, while cold-temperate and temperate forests, which account for most of China's forests and available forestry land, would decrease.

GHG SOURCES AND SINKS INVENTORY FOR CHINA, 1990

In 1990, global anthropogenic CO_2 emissions were nearly 5.7 billion tons of carbon, of which China accounted for about one-tenth. Per capita CO_2 emissions from China were 0.6 tons of carbon (tC), compared with the United States (5.3 tC), Japan (2.3 tC), and the former Soviet Union (3.7 tC).

An inventory of major emissions and sinks of GHGs has been made for China using the methodology adopted by the IPCC.[5] The benefit of using this standardized methodology is that it allows countries to prepare emissions inventories that are comparable; the drawback is that the emission coefficients may not reflect the true situation in some countries. While IPCC coefficients have been used throughout this report, major discrepancies that are believed to exist with the Chinese situation are discussed.

The GHGs that are estimated in this report are carbon dioxide (CO_2), methane (CH_4), and nitrous oxide (N_2O). In order to compare these gases, their "global warming potential" (GWP) has been used.[6] Chlorofluorocarbons (CFCs), which in addition to de-

Table 1.2 CO_2 emissions from major contributors, 1990 (mtC)[a]

	CO_2	Percent
China	596[b]	10
United States	1,222	21
Japan	248	4
USSR (former)	1,034	18
Europe	1,198	21
Other	1,392	22
World	5,690	100

[a] All weights are expressed in terms of the molecular weight of carbon (C), as opposed to the full molecular weight of CO_2.

[b] The estimate for China is slightly different from the one calculated in this report. For comparative purposes, the Chinese estimate for this table was not modified.

Source: World Resources, 1990-1991, World Resources Institute, (Washington, DC: 1992).

stroying the ozone layer are potent greenhouse gases, have not been estimated for this report since they are currently a relatively small portion of China's total emissions and since China is phasing out their use under the Montreal Protocol Convention.[7]

Energy consumption is the source of 82 percent of China's GHG emissions. Methane accounted for approximately 13 percent of GHG emissions in 1990

---

[5] The emission estimation methodology used in this study is based on the results of the IPCC-commissioned Organization of Economic Cooperation and Development (OECD) Experts Meeting held in Paris in February 1991. The methodology outlined at this meeting has been adopted and distributed by the IPCC. *Estimation of Greenhouse Gas Emissions and Sinks, Final Report from the OECD Experts Meeting, 18-21 February 1991*, prepared for the IPCC, August 1991.

[6] The GWPs used here are for a 100-year average: methane, 11 times CO_2; and nitrous oxide, 290 times CO_2. It should be noted that the GWP for methane of 11 times CO_2 is for direct effects only. The 1990 IPCC scientific assessment estimated that the total GWP of methane, including both direct and indirect effects, was 21 times CO_2. It is likely that a 1995 IPCC supplement will recommend the use of a GWP for methane of 25 times CO_2. See John T. Houghton (ed.) *Climate Change 1992, The Supplementary Report to the IPCC Scientific Assessment*. (Cambridge: Cambridge University Press, 1992).

[7] CFCs have been estimated in other recent climate change studies for China. See *National Response Strategy for Global Climate Change: People's Republic of China*, Asian Development Bank, Chinese State Science and Technology Commission, Final Report, September 1994.

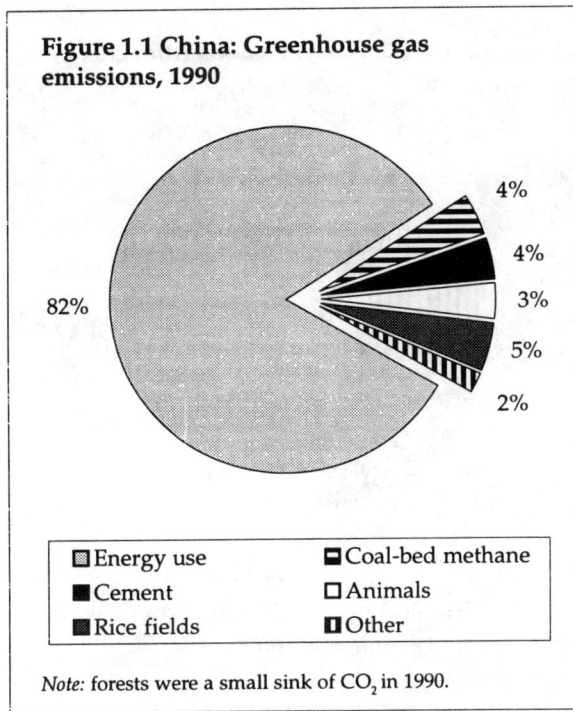

Figure 1.1 China: Greenhouse gas emissions, 1990

82%

4%
4%
3%
5%
2%

Legend:
- Energy use
- Cement
- Rice fields
- Coal-bed methane
- Animals
- Other

Note: forests were a small sink of CO_2 in 1990.

Energy consumption

China has an abundance of coal and a relative lack of petroleum and natural gas reserves; thus, coal is the major source of CO_2 emissions in China. Unlike developed countries, where coal is used mainly in power generation, in China the power sector accounts for only about a quarter of total coal consumption. Most coal is consumed directly by industry for steam generation and by the residential sector for cooking and heating.

The industrial sector accounts for nearly three-quarters of China's CO_2 emissions from energy consumption. Within industry, the largest energy-consuming sectors are electric power, building materials, iron and steel, chemicals, and the energy production industry itself—coal mining, coal processing, and oil and gas extraction. Together, these five sectors account for 79 percent of CO_2 emissions from the industrial sector.

Industrial boilers used outside the power sector consumed more than 350 million tons of coal in China in 1990, accounting for about 35 percent of the country's coal use and about 30 percent of GHG emissions from energy consumption. There are an estimated 430,000 industrial boilers in China, of which more than 95 percent are coal-fired and very small by international standards.

China uses much less energy for transportation than other developing and developed countries. When correcting for statistical differences in the way transport energy use is reported in China, transport accounts for about 7 percent of total energy use.

The residential sector accounts for about 14 percent of commercial energy consumption, most of which is coal used for cooking and space heating. Non-commercial biomass fuels, including quality fuelwood, crop residues, and some animal dung, amounted to approximately 300 million tons of coal equivalent (mtce) in 1990. CO_2 emissions from biomass burning not obtained through deforestation are ignored, since CO_2 will be recaptured in subsequent plant growth. Non-CO_2 emissions from biomass fuels, such as CH_4 and N_2O, are, however, included in total GHG emissions.

Recent Chinese studies indicate that a high percentage of carbon remains unburned in China because of inefficient fuel combustion in small industrial and commercial boilers and in residential stoves.[9] Based

(CO_2 equivalent) with coal mining, rice fields, ruminant animals, and animal wastes contributing 88 percent of methane emissions. All other sources of emissions, including CO_2 from cement manufacturing, methane from landfills, N_2O from fertilizer, and non-CO_2 emissions from forests and land use changes, accounted for about 6 percent of China's total GHG emissions in 1990.[8]

There is a high degree of uncertainty in GHG emissions estimates depending on the nature of emissions (such as deforestation) and the difficulty of measurement (such as methane from rice and nitrous oxide from fertilizer use). Some estimates, however, are fairly accurate. For example, the range of uncertainty for CO_2 emissions from fossil fuels and cement manufacture is approximately 10 percent. In contrast, the range of uncertainty for deforestation and land use change is conservatively in the range of ±60 percent and there is even greater uncertainty for most sources of methane and nitrous oxide; the range of estimates for many nitrogen fertilizer coefficients is 0.001 to 6.84 (percent N_2O-N produced).

---

[8] For a complete discussion of the estimation methodology and results for China, see the project subreport, *Estimation of Greenhouse Gas Emissions and Sinks in China, 1990*, August 1994.

on these studies, Chinese researchers have estimated that CO_2 emissions from coal consumption in China may be overstated by as much as 10 percent if IPCC coefficients are used. As fuel combustion efficiency improves over time, the true coefficient in China will increase. Therefore, the joint study team concludes that using the international coefficient represents an upper bound of CO_2 emissions from coal consumption in China in 1990.

Energy production, storage, and distribution

Methane is the primary greenhouse gas emitted in energy production. It is released in the process of oil and gas production, oil refining, and natural gas transmission and distribution. The largest source of methane from energy production in China is coal mines; methane must often be vented from underground mines for safety reasons. Because of the large quantity of coal that is mined in China, and the fact that more than 95 percent of the coal comes from underground mines, methane emissions from this source are estimated to be quite high. Based on the IPCC methodology and partial data from Chinese coal mines, methane emissions from coal mining in China are estimated to be 12–18 mt per year. According to estimates by Chinese coal production experts, methane emissions from coal production in 1990 were between 4 and 7 mt per year. Because of the difference in these estimates, an average of the two—10.67 mt—was used for this report.

Agricultural sector emissions

Methane production from rice fields is a function of the age of the rice plants, the length of time the fields are flooded, and the amount of decomposing biomass on the fields. For rice cultivation, the estimated daily emissions flux of 0.19–0.69 g CH_4/m^2 derived by researchers in Hangzhou has been used to calculate emissions. Methane emissions from rice were estimated to be 11.0–12.8 mt in 1990.

Ruminant animals—yellow cattle, buffalo, yak, camels, and goats—produce methane in digesting food; ruminants account for about 95 percent of CH_4 emissions from domesticated animals. An additional

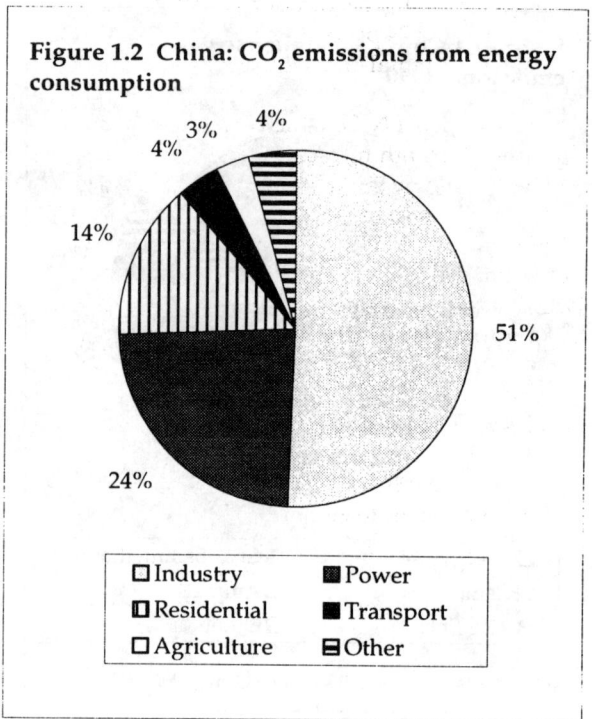

Figure 1.2 China: CO_2 emissions from energy consumption

- Industry — 51%
- Power
- Residential
- Transport
- Agriculture
- Other

(24%, 14%, 4%, 3%, 4%)

Table 1.3 China: Greenhouse gas emissions, 1990

	CO_2(mtC)	CH_4(mt)	N_2O(mt)
Energy consumption	650	0.06	0.18
Oil and gas leakage	...	0.18	...
Coal-bed methane	...	10.67	...
Cement	29	...	...
Landfill gas	...	0.79	...
Rice fields	...	11.90	...
Animals: enteric fertilization	...	6.24	...
Animal wastes	...	2.70	...
Fertilizer	...	...	0.03
Biomass burning	...	3.62	0.05
Total	679	36.16	0.26

Source: Estimation of Greenhouse Gas Emissions and Sinks in China, August, 1994 subreport prepared for China GHG Study.

---

9 In some parts of China, so much carbon is left in the ash that the coal ash is sometimes reused as a fuel or combusts spontaneously. To the extent that the carbon in coal ash is combusted, the resulting CO_2 emissions would raise China's emission coefficient toward international levels.

2.7 mt of methane was produced from animal wastes in 1990. Together, domesticated animals account for about 24 percent of methane emissions and 3 percent of total GHG emissions.

The emission of N_2O, which is produced naturally in soil through nitrification and denitrification, is enhanced by the application of nitrogen fertilizers. N_2O is also released through commercial energy consumption and biomass burning. Although the range of emissions estimates for N_2O is large, the high estimate from all sources amounts to 2.5 percent of total GHG emissions in China in 1990.

Cement

There are several non-energy related industrial activities that generate greenhouse gas emissions, the most notable of which is CO_2 from cement production. In 1990, cement production in China was about 210 million tons, resulting in the emission of 29 mtC—about 4 percent of China's total GHG emissions in 1990.

Forests and land use changes

There is much uncertainty associated with forestry sector emissions. Nonetheless, based on a review of land use in China, specifically the level of afforestation and the conversion of forestry lands to other uses, forests are estimated to have been a net sink of atmospheric carbon in 1990. The amount of carbon sequestered in woody biomass and soils in China is estimated to be about 7 mtC, which reduced total GHG emissions in 1990 by about 1 percent.

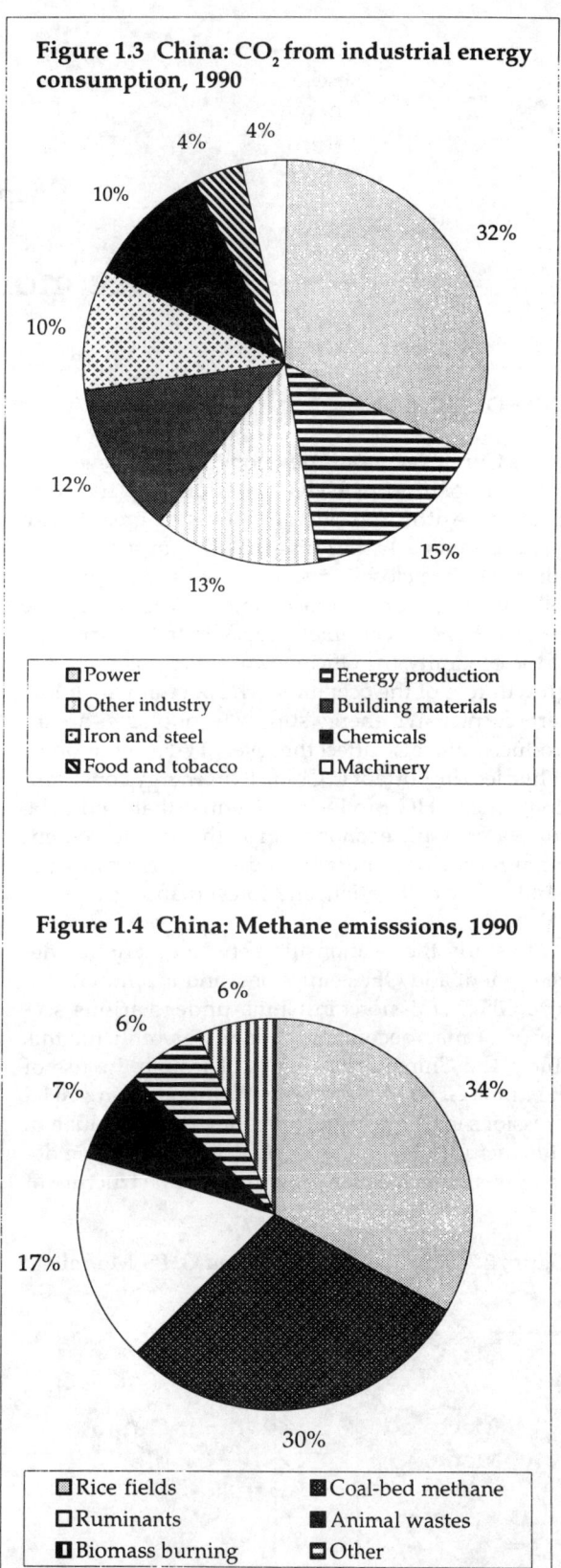

Figure 1.3 China: CO_2 from industrial energy consumption, 1990

Power, Energy production, Other industry, Building materials, Iron and steel, Chemicals, Food and tobacco, Machinery

Figure 1.4 China: Methane emisssions, 1990

Rice fields, Coal-bed methane, Ruminants, Animal wastes, Biomass burning, Other

Chapter 2

Macroeconomic Analysis

INTRODUCTION

As China's economy develops, both the level and the composition of GHG emissions are likely to change. Anthropogenic emissions of greenhouse gases, particularly those associated with energy consumption, are closely correlated with economic development. At the macroeconomic level, energy use and its change over time are related to the structure of the economy, the efficiency of energy use, and the growth rate of the economy. The pace at which low carbon-intensive energy supply technologies are introduced will also affect the level of GHG emissions. While less important in China than energy consumption, other GHG-producing activities that tend to be correlated with economic growth include cement manufacturing, energy production, animal husbandry, rice cultivation, and forest planting.

To study the relationship between economic development and GHG emissions, and to simulate future GHG emissions in China under various scenarios, a macroeconomic model was built for this study: the China GHG Model. Through the use of the China GHG Model, the key factors affecting GHG emissions in China in the future have been evaluated. This includes factors associated with economic development, such as the growth rate and structure of the economy, and factors appropriate for targeted policy and investment intervention, such as energy efficiency and afforestation. This chapter addresses the magnitude of GHG emissions associated with different factors, while the costs of reducing emissions are assessed in Chapter 3.

THE CHINA GREENHOUSE GAS MODEL

The China GHG Model is used to simulate economic growth, structural change, energy consumption, and the resulting emission of GHGs and selected local air pollutants within China over the next twenty-five years. The impacts of alternative energy, afforestation, and agricultural programs have been evaluated separately and the results incorporated into the China GHG Model. The model has four main components: i) a macroeconomic model, ii) an input-output model, iii) energy coefficients, and iv) emissions coefficients (Figure 2.1).

Macro model. The macro model is used to project final demand (Y): the demand for goods and services, investment, government expenditures, and net exports. The macro model used for this study was originally built as part of a collaborative project between Stanford University, the University of Penn-

Figure 2.1 Schematic of the China GHG Model

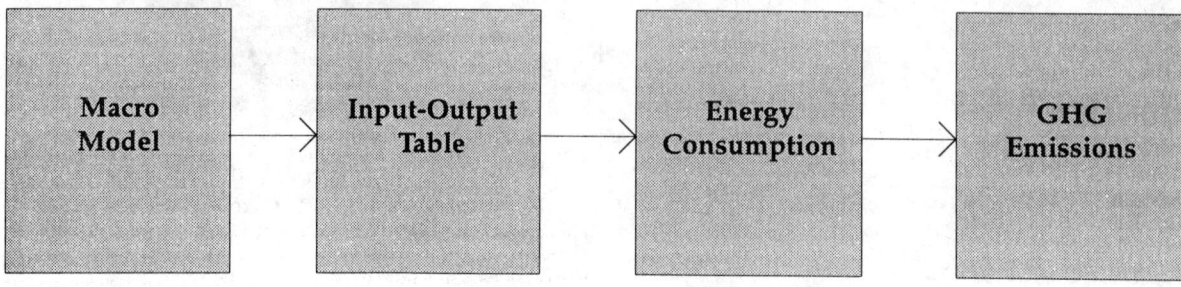

sylvania, and the Chinese Academy of Social Sciences. It has been modified for this study by including energy as an additional factor of production and by separating the energy sector from heavy industry. The model is econometric, consisting of roughly 250 equations, 140 of which are regression equations based on Chinese economic and social statistics (1965-1992). Together, the equations describe the Chinese economy, such as the production of goods, the population and labor force, income determination, consumption, capital formation, public finance, money and banking, prices, international trade, and the balance of payments.

Input-output model. An input-output (I-O) table represents the structure of the economy and the efficiency with which goods and services are produced. The I-O coefficients, referred to as the "A" matrix, represent the inputs required from each sector to produce a unit of output. For example, to produce a unit of steel may require inputs from the coal, heavy machinery, and metallurgy sectors. Once the A matrix has been determined, it is possible to estimate the total inputs (X) required by the economy—such as labor, energy, steel, and cement—to produce a given mix of final demands (Y). For this study, China's 1987 input-output table was aggregated into eighteen sectors and updated to the base year 1990 using both the 1987 and 1981 Chinese I-O tables.[1] To use I-O analysis for future projections, it is necessary to estimate how the I-O coefficients change over time. I-O coefficients for China were projected to the years 2000, 2010, and 2020 using Chinese historical I-O tables and cross-sectoral analyses of Chinese provincial I-O tables, and by reviewing historical changes in I-O coefficients for other countries, including Japan (1965, 1970, 1975, 1985), the United States (1939, 1947, 1958), West Germany (1965, 1970, 1975), and the United Kingdom (1965, 1970, 1975).

Energy coefficients. The energy coefficients in the model represent the amount of coal, petroleum, natural gas, and electricity needed to produce a unit of output for each of the eighteen sectors. The technical energy coefficients in the I-O table have an enormous effect on the amount of energy required to produce a given level of economic output. Given the importance of the energy coefficients for GHG emissions, a bottom-up approach, based on a spreadsheet of product-specific energy coefficients, was used to strengthen and corroborate the results obtained from historical and cross-country comparisons of energy I-O coefficients. The joint study team conducted a detailed analysis of current and future energy consumption, estimating the energy consumption requirements for producing major commodities within each of the eighteen sectors. Four factors were considered in modifying the energy coefficients: (i) the growth of the economy, since this affects the rates of structural change and investment in new energy-efficient capital stock; (ii) the scale of new plants and equipment relative to existing facilities; (iii) the rate of adoption of new energy-saving technologies and industrial processes; and (iv) increases in product diversity and improvements in product quality, which affect the amount of energy per unit of output value. By varying the product-specific energy efficiency coefficients, it is possible to assess the impact of improvements in specific technologies on energy use and emissions.

Emissions coefficients. The China GHG Model generates emissions matrices of both global and local air pollutants for each level of economic output. The primary GHG estimated by the model is CO_2 from energy consumption. The model assumes that the percentage of carbon oxidized for each fuel is the same in all fuel combustion applications; thus, once the quantity of energy consumed is determined, the amount of CO_2 emitted can be calculated. Local pollutants estimated by the model include total suspended particulates (TSP) and sulfur dioxide (SO_2). For local pollutants, emission factors were estimated for each of the eighteen sectors based on the average ash and sulfur content in the fuel consumed by that sector, and on the emission rate and current level of control within that sector.[2] In addition to CO_2 emissions from energy consumption, the China GHG Model also estimates CO_2 emissions from cement, and

---

[1] This work was carried out under the direction of Professor Chen Xikang, Institute of Systems Science, Chinese Academy of Sciences. Professor Chen is a leading expert in the field of input-output analysis and was responsible for building China's national input-output tables in the 1970s. He presently forecasts economic indicators for China, such as national grain production, using I-O analysis.

[2] The sectoral emission coefficients for 1990 were estimated by Chinese energy and environmental specialists from line ministries and were reviewed and modified by experts from the National Environmental Protection Agency. Because the mandate of this project is to look at global pollutants, the current version of the China GHG Model does not take into account improvements in TSP and SO_2 control by new equipment or the adoption of specific control devices for TSP or SO_2. As such, the emission levels of SO_2 and TSP for future years can be regarded as a worst-case scenario. These deficiencies are expected to be corrected in later versions of the model.

CH_4 emissions from rice fields, coal mining, rice, and animal husbandry. Together, these sources of GHGs accounted for approximately 98 percent of China's total GHG emissions in 1990. Forestry emissions and sequestration are generated "off-line" in a separate forestry model (see pp. 51–53).

BASELINE GHG SCENARIO

For purposes of comparison, a baseline scenario of GHG emissions for China for the years 2000, 2010, and 2020 was generated. The Baseline GHG Scenario assumes that the current trends in economic growth in China will continue, paralleling the development of other East Asian economies over the past thirty years, and that the energy efficiency improvements that have been made in China over the past decade will also continue. Under the baseline GHG scenario, per capita gross domestic product (GDP) in China reaches US$2,600 in 2020 (1990 constant dollars), up from roughly $370 in 1990; total energy consumption surpasses 3.3 billion tce, up from 1 btce in 1990, and 1.18 times the level in the United States in 1990; and total GHG emissions increase threefold, surpassing 2.3 billion tons carbon (tC).[3] The results and assumptions used for the Baseline GHG Scenario are shown in Table 2.1 and described in detail below.

ECONOMIC DEVELOPMENT AND GHG EMISSIONS

Future GHG emissions from China will depend primarily on the interrelated issues of economic growth, the structure of the economy, and the structure of energy use. In contrast to the specific policy options for reducing GHG emissions reviewed in the next section, these factors are largely outside the direct influence of government policy.[4] Nonetheless, sensitivity analysis shows that macroeconomic and structural issues are the most important determinants

---

[3] All income and cost data presented in this chapter are in 1990 constant prices. In addition, Chinese yuan are converted to US dollars using the official 1990 exchange rate of Y4.7/US$.

[4] Measures that can be taken to improve the overall efficiency of the economy, and thus reduce GHG emissions, are discussed on pp. 57-59.

Table 2.1 Baseline GHG Scenario: Results and assumptions for 2020

Results

Greenhouse gas emissions:

Total	2,398 mtC and equivalent (threefold increase over 1990)
From energy consumption	2,089 (87%)
Other	309 (13%)
Per capita	1.6 tC
Energy consumption	3,300 mtce
SO_2 and TSP emissions	55 mt, 48 mt, respectively

Assumptions

Economic growth	Average 8 perent per annum between 1990 and 2020.
Macroeconomic structure	Contribution to GDP: slight decline of industry, growth of services, decline in agriculture.
Energy efficiency	Continuation of energy efficiency improvements. Energy-GDP elasticity of 0.5–0.6 maintained between 1990 and 2020.
Alternative energy	Expansion of hydro, nuclear, and renewable programs. Alternatives account for 22 percent of total electricity supply in 2020 and 8 percent of total energy supply.
Forestry	No change in the absolute level of net afforestation from 1990.
Agriculture	No change in GHG-generating emission factors (rice, animals, fertilizer) from 1990.

Source: Joint Chinese-international study team, China GHG Study.

of future GHG emissions in China.

Economic growth

Two scenarios of future income growth in China were evaluated using the China GHG Model. The *high-growth* scenario assumes that there are few political or social disruptions affecting the economy, the relatively high savings rate is maintained, and the economic reform program continues. Although the high-growth scenario is optimistic in its assumptions, many experts have predicted that the growth rate in China over the next twenty-five years will be even higher. A *slower-growth* scenario was also evaluated, representing the lower bound of a stable, though less robust, Chinese economy.

High growth. In the high-growth scenario, GDP grows by an average of 8 percent per year between 1990 and 2020. The growth rate during the first decade (1990–2000) is assumed to be 9.5 percent; during the second (2000–2010), 8.0 percent; and during the third (2010–2020), 6.5 percent. Actual income growth in China was 6.8 percent per year between 1965 and 1980, and 9.6 percent per year during the 1980s. Given China's double-digit growth in the early 1990s, its economy would have to grow at less than 8.5 percent annually for the remainder of the decade to fall short of the high-growth scenario.

In the high-growth scenario, China's GDP in the year 2020 reaches US$3.8 trillion (1990); roughly 1.4 times the size of Japan's economy and nearly three-quarters the size of the U.S. economy in 1990. Per capita income in 2020 in the high-growth scenario would be approximately US$2,600 (1990), comparable to income levels at the lower end of "upper-middle-income" countries in 1990, such as Brazil, South Africa, Mexico, Venezuela, and Hungary.[5]

Average growth exceeding 8 percent annually over several decades is not unprecedented among East Asian economies. Korea's economy grew at approximately 10 percent per year between 1965 and 1990, while Taiwan's GDP increased nearly 9 percent per year between 1951 and 1991. The Japanese economy grew an average of 8.7 percent between 1946 and 1976, and its annual GDP growth was more than 10 percent between 1950 and 1970.

[5] Income data and definitions are from the World Bank, *World Development Report, 1992* (New York: Oxford University Press 1993), pp. 218-219.

Figure 2.2 GDP comparison: Japan and U.S. (1990) and China high-growth

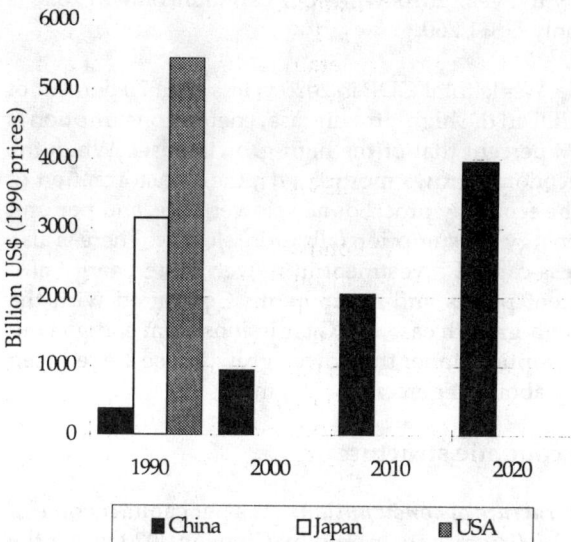

Sources: China Statistical Yearbook; World Development Report, 1992; China GHG Model, joint study team.

There are several factors specific to China's economy that are conducive to high growth in the coming decades. The economic reform program, which is likely to continue and deepen, will create new opportunities and incentives for expanding the domestic market and international trade. China's high savings rate, which has increased since the 1980s, is essential for providing investment funds to continue the economic expansion. In terms of age structure, China's population is very young, which not only increases the level of savings, but also provides an ample pool of workers and consumers. Japan, Korea, and Singapore all had similar population age structures during their economic booms. The sheer size of China's economy represents an almost limitless market for goods and services, which means that China will not need to rely on export-led growth as have other East Asian economies. For these reasons, the high-growth GDP scenario was used in the Baseline GHG Scenario outlined in Table 2.1.

Slower growth. Whether caused by government policy or market forces, slower economic growth results in lower energy consumption and GHG emissions. In the slower-growth case, China's economy is assumed to grow 1 percent less than in the high-growth case during the 1990s and 1.5 percent less for the decades between 2000 and 2020. Total GDP in the slower-growth case reaches only US$2.5 trillion by the year 2020, while per capita income in 2020 is only US$1,760.

While total GDP in 2020 is less than 70 percent of GDP in the high-growth case, energy consumption is 90 percent that of the high-growth case. When the economy grows more slowly, the transformation of the economy proceeds at a slower pace, and per unit energy consumption falls more slowly. There is also less capital investment in newer, more energy-efficient plants and equipment. Compared with the high-growth case, GHG emissions from energy consumption under the slower-growth case are reduced by about 14 percent, or 290 mtC.

Economic structure

Structure of consumption. At a per capita income of $2,600, which is reached by China in 2020 under the high-growth income case, material goods typically dominate household consumption. The model predicts that by the year 2020, the share of agricultural products (grain, vegetables, meat, and fish) in final demand will fall for household consumers, while the share of services (public utilities, health care, education, and housing rental) will increase. Consumption of light industrial products as a share of total consumption will remain roughly the same between 1990 and 2020.

Although the model cannot provide details on the consumption of specific commodities, it is likely that the composition of light industrial products would shift away from basic clothing and processed foodstuffs toward more high-quality consumer goods. If the distribution of income in 2020 remains roughly the same as in 1990, some upper-income households will be able to afford automobiles and a range of high-valued services. However, at a per capita income of less than $3,000, the average Chinese household will still concentrate its spending on material goods such as household appliances (e.g., refrigerators, washing machines, and air conditioners) and housewares (e.g., furniture and carpeting).

Structure of the macroeconomy. The structure of China's economy in 2020 will be largely determined by the pattern of household consumption. As shown in Figure 2.3, little change is expected in the share of heavy and light industry in total economic output over the next three decades. Industrial growth must remain strong for China to satisfy consumer demand

Table 2.2 Structure of consumption: Share of household expenditures

	1990 (%)	2000 (%)	2010 (%)	2020 (%)
Agricultural products	34	21	15	11
Light industrial products	40	45	41	37
Services	14	21	29	35

Source: China GHG Model, joint study team.

for light industrial products, for construction, and for the manufacturing sector's demand for heavy industrial producer goods. The share of the tertiary or services sector, however, is expected to increase sharply, largely offsetting a fall in agriculture's contribution to GDP. The overall effect of this macroeconomic structural change on GHG emissions is not large, because neither agriculture nor services are energy-intensive sectors. As shown in the following section, however, other structural changes, especially within the industrial sector, have a major bearing on future GHG emissions.

A baseline energy use scenario

Future energy consumption patterns are the most important determinant of China's future GHG emissions. The level of energy use required to sustain China's projected economic growth is uncertain, however. As in other countries, there is uncertainty about China's ability to improve the technical efficiency of energy use in various applications and the level of energy services that will be demanded by its population. In China's case, structural changes, particularly in industry, will determine the energy intensity of economic output. In the high-growth GDP case, net industrial output is projected to grow ninefold between 1990 and 2020. How efficient will China's new industries be in converting a variety of intermediate inputs into value added? How much will the productivity of industrial manufacturing equipment increase? How will the industrial product mix change? How will improvements in product quality and specialization affect energy use per unit of industrial value added? Structural changes stemming from such factors were the main reason for China's low energy consumption and GDP growth elasticities during the last fifteen years, and further structural

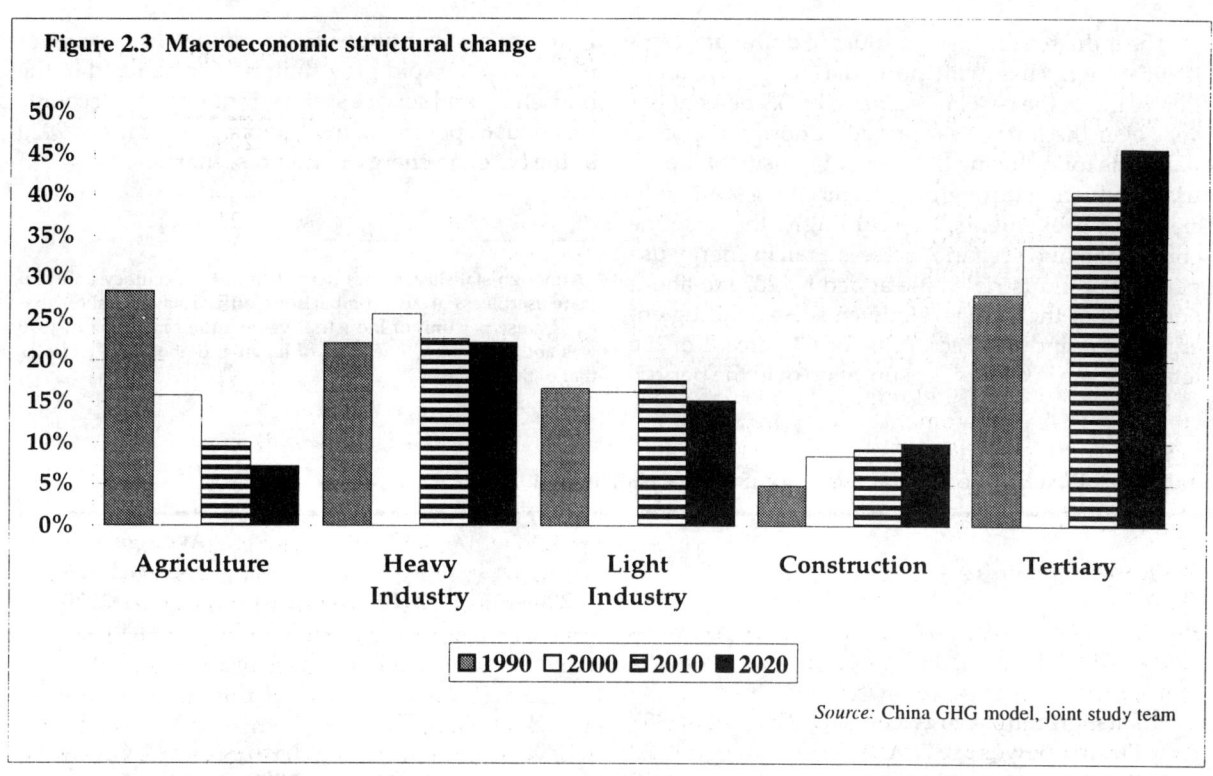

Figure 2.3 Macroeconomic structural change

Source: China GHG model, joint study team

Legend: ■1990 □2000 ▤2010 ■2020

changes are expected to continue to be the main factor in reducing the energy intensity of China's economy in the future.[6]

For reference, the joint study team prepared a detailed "baseline case" energy demand scenario to assess the relative importance of different factors and potential interventions on total GHG emissions. This case is intended to model how energy demand may evolve under "business-as-usual" conditions. Structural changes in the economy continue to reduce energy intensity (pp. 23-24), and continued progress is made in improving the technical efficiency of energy use in key applications. Assuming the high-growth GDP scenario, the baseline energy demand scenario shows growth in primary commercial energy consumption of about 3.3 times between 1991 and 2020, and an average annual growth rate of 4.1 percent (Table 2.3). The energy use/GDP growth elasticity during this period is about 0.5. This compares with elasticities of 0.52 from 1981 to 1990, and less than 0.4 from 1991 to 1993.

Coal demand roughly triples between 1991 and 2020 in the baseline case, despite major declines in the energy intensity of the economy and faster growth in the use of other fuels. The baseline case envisages growth in natural gas use of about 7 percent per year, which would require major new discoveries or substantial imports or both. Primary power production, predominantly hydro and nuclear power, would also increase relatively quickly. Accordingly, the share of coal in total commercial energy use would fall from 76 percent in 1990 to 67 percent in 2020. Even so, massive requirements for power generation and industrial process heat and steam for the country's economic expansion continue to drive coal demand upward, with tremendous logistical and environmental implications. Power sector demand for coal rises from some 250 million tons in 1990 to about 1,300 million tons in 2020, and the proportion of coal used for power generation would rise sharply, from 24 percent in 1990 to 42 percent in 2020. Non-power industrial coal demand rises from 546 million tons in 1990 to more than 1,400 million tons in 2020. Owing to relatively modest growth in household cooking and heating demand, energy efficiency gains, and some substitution of other fuels, coal demand in the household sector is expected to remain fairly constant, thus reducing the share of household consumption in total coal use.

[6] See *China: Energy Conservation Study*, World Bank Report No. 10813-CHA, February, 1993 (Washington, DC:1993).

The industrial sector continues to dominate commercial energy use, with industrial energy consumption rising in the baseline scenario by 3.9 percent per year over the thirty-year period. Enormous potential exists for reducing the energy intensity of industrial production through structural changes and efficiency improvements.[7] Accordingly, the baseline energy demand scenario foresees a fall in energy use per unit of industrial value added by 2020 to about one-third of the 1990 level. Even so, expanding energy use in industry accounts for 63 percent of the total increase in energy consumption over the period.

Industry also continues to dominate electric power use. Despite rapid growth in power demand in the household and service sectors, industry still accounts for about 65 percent of total electric power use in 2020 in the baseline energy demand scenario.

---

[7] Although statistical issues detract from the accuracy and ultimate usefulness of such comparisons, official statistics show that energy use per unit of industrial value added in China in 1990 was about eighteen times that of Japan and about eleven times that of South Korea.

Table 2.3 Baseline commercial energy use scenario, 1990-2020

	1990	2000	2010	2020	Average annual growth rate, 1991-2020
Total primary energy use (mtce)	987	1,560	2,380	3,300	4.1
Electric power use (TWh)	621	1,300	2,430	3,850	6.3
Per capita energy use (kgce)	863	1,200	1,700	2,280	3.3

Primary energy use by type	1990			2020	Average annual growth rate, 1991-2020
Coal (million tons)	1,053			3,100	3.7
Oil (million tons)	112			440	4.7
Natural gas (billion m³)	15			115	7.0
Electric power (TWh)	126			870	6.7

Coal use by sector	1990		2020		Average annual growth rate, 1991-2020
	mt	%	mt	%	%/year
Power	251	24	1,300	42	5.7
Non-power industry:	546	52	1,440	46	3.3
Ferrous metals	103	10	215	7	...
Cement	50	5	155	5	...
Other building materials	99	9	295	10	...
Light industry	116	11	300	10	...
Households	167	16	165	5	0
Other	89	8	195	6	2.6
TOTAL	1,053	100	3,100	100	3.7

Sources: China Energy Statistical Yearbook, 1991; China GHG Model.

Per capita commercial energy consumption in the baseline energy demand scenario increases from 863 kgce in 1990 to about 2,280 kgce in 2020, an annual growth rate of 3.3 percent (Figure 2.4). In the past, China's per capita commercial energy consumption has been high compared to other low-income developing countries. Under the baseline scenario, per capita energy consumption levels relative to per capita GDP approximate international trends. By 2020, China's per capita energy use would be within the mainstream of international experience among countries with similar income levels. In 2010 and 2020, per capita energy use would be less than half that of some Eastern European countries at a similar income level, but substantially higher than several other large developing countries. In 2020, per capita energy consumption would be less than the average for "upper-middle-income" countries in 1990 and less than one-quarter of that in the United States in 1990.

Sources of future declines in energy intensity

In 1990, China's economy was among the most energy-intensive in the world, registering a level of energy use per unit of GDP of between three and ten times that of major developed countries. In the baseline energy demand scenario, the energy intensity of China's economy falls from about 2.7 mtce per US$1,000 of GDP in 1990 to about 0.9 mtce per US$1,000 of GDP, in comparable prices, in 2020.

China's energy intensity is several times higher than that of most other large countries because of its unique economic structure. Similarly, the decline in energy intensity foreseen in the baseline case is primarily because of structural factors. As shown on page 26, there is tremendous scope for further improvement in the technical efficiency of energy use; that is, the efficiency of energy use per physical unit of output. Even so, technical efficiency improvements still amount to only 15–25 percent of the estimated total decline in energy intensity.[8] Structural factors, as described below, would account for 75–85 percent of the total energy intensity decline.[9]

---

[8] Estimates of the decline in energy intensity by source are presented in a subreport prepared for this project. See *Energy Demand in China: Overview Report*, February 1995. Forthcoming.

[9] Additional analysis of these structural factors in China can be found in *China: Energy Conservation Study*, World Bank Report No. 10813-CHA, February 1993 (Washington, DC:1993).

Figure 2.4 Per capita GDP and energy consumption, low and middle income countries, 1990 and China baseline energy demand scenario

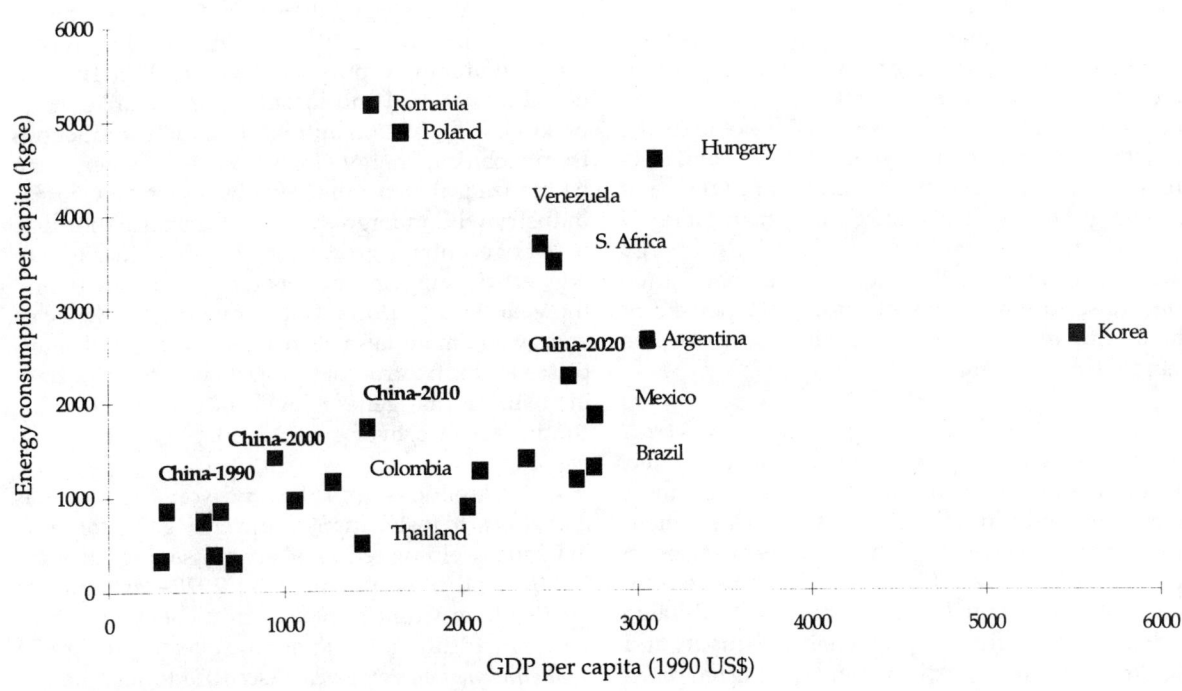

Changes in macroeconomic structure. One of the reasons China's energy intensity is higher than most other countries with similar per capita income levels is that industry currently makes up about 44 percent of China's GDP, a share much higher than that of other low-income countries and even higher than that of Japan. This drives up energy intensity because industry is far more energy intensive than the other major sectors. However, the projected slight decline in the share of industry by 2020 will provide some energy savings, amounting to about 7 percent of the energy intensity decline in the baseline case. Moreover, the increasing share of industry in GDP, which tends to drive up energy use/GDP growth elasticities in other countries as they proceed from low- to middle-income levels, will not occur in China.

Residential energy use. While GDP increases tenfold, residential energy use in the baseline energy demand scenario would not quite triple over the thirty-year period, even with an increase in residential electricity demand of more than ten times. This creates a strong downward pressure on energy intensity. Technical energy efficiency gains in this sector are an important factor, but over the long term, the demand for household cooking and heating increases more slowly than GDP as rapid economic development proceeds. This structural effect is estimated to account for 10–15 percent of the total energy intensity decline.

Changes in the shares of industrial subsectors in industrial output. Changes in the contribution of different industrial subsectors to industrial output lead to changes in overall energy intensity, since energy intensity varies among industries. For example, the growth of less energy-intensive machinery and electronics industries, and the declining importance of the energy-intensive metallurgy and chemical fertilizer industries will contribute to a decline in energy intensity. Changes in the output mix among industrial subsectors will account for about 9 percent of the decline in total energy intensity in the baseline energy demand scenario.

Changes in the product mix within industrial subsectors. A shift in the product mix within different industrial subsectors has been the single most important factor in reducing China's energy intensity over the last decade. This trend is expected to continue, especially with further economic reform. Much of the expected increase in industrial value added will come from improvements in quality and the production of more diverse and specialized products, rather than from the production of energy- and raw-material intensive basic industrial goods. In the textile and garment industry, for example, substantial growth in value added will come from increases in product value, in terms of quality and fashion, rather than large increases in the amount of cloth or number of shirts produced. Declines in energy use per unit of output value resulting from this trend will be especially important in the chemical, machinery, building materials, and light industry sectors, which together accounted for more than 60 percent of industrial final energy use in 1990. Although it is hard to quantify the future importance of this factor, it is estimated to provide 35–40 percent of the total energy intensity decline in the baseline energy demand scenario.

Other structural changes. The net effect of structural changes in the non-industrial sectors, such as agriculture, commerce, transportation, and other services, is estimated to account for a final 10–15 percent of the total decline in energy intensity.

OPTIONS FOR LIMITING GHG EMISSIONS

Energy efficiency

Although structural effects may have the greatest impact on the energy intensity of China's economy, technical efficiency levels—energy consumption per physical unit of output—can be more directly influenced by national and local policies. Unlike many other GHG reduction options, emissions reductions from technical energy efficiency improvements can be realized almost immediately. Because Chinese industry will undergo dramatic transformation as a result of economic growth, the opportunities for energy efficiency gains are greatest between now and the year 2010. Efforts to improve the efficiency of energy and materials use in all new industrial enterprises and infrastructure projects will be important in ensuring that general levels of energy efficiency continue to improve beyond 2010.

The baseline energy demand scenario assumes that efforts to foster energy-efficiency gains continue in China, yielding technical energy savings of some 1,000–1,700 mtce per year by 2020—an amount greater than China's total energy consumption in 1990. In addition, the joint study team prepared a *high-efficiency energy demand scenario* to test the potential impact of additional improvements in technical energy-efficiency levels. In this scenario, key unit consumption parameters were adjusted, primarily for the industrial sector, to reach levels more closely approximating advanced international rates. Economies of scale in the production of energy-intensive products are a large part of the additional energy savings under the high-efficiency scenario. Table 2.4 lists and compares the most important assumptions for both demand scenarios.

In the high-efficiency scenario, total primary commercial energy demand in 2020 is 2,840 mtce. This represents an additional energy savings of 460 mtce per year and a reduction in GHG emissions of 330 mtC compared with the baseline energy demand scenario.

IMPROVEMENTS IN INDUSTRIAL ENERGY EFFICIENCY

In general, there are three types of industrial investment projects that can yield gains in energy efficiency:

1. *industrial modernization,* where major restructuring of the enterprise results in energy savings and other benefits;

2. *industrial energy conservation projects,* such as recovery of waste heat, where the primary goal is energy savings; and

3. *improvements in widely used industrial equipment,* such as electric motors and industrial boilers.

Industrial modernization projects. Although efforts at energy efficiency often focus on specific energy-saving technologies, industrial modernization has had the greatest impact on energy savings in China in the past decade and will continue to for several decades. There are three aspects of industrial modernization that are most important for energy effi-

Table 2.4 Comparison of selected unit energy consumption levels, baseline and high-efficiency (HE) scenarios, 1990-2020

		China			Japan
		1990	2020 (baseline)	2020 (HE)	1980
Steel[a]	(kgce/t)	1,610	1,284	857	...
Cement	(kgce/t)	208	196	135	135
Ammonia	(kgce/t)	2,066	1,665	1,258	1,000
Thermal power	(kgce/MWh)	427	348	345	338
Caustic soda	(kgce/t)	1,790	1,325	1,000	1,000
Ethylene	(kgce/t)	1,580	1,450	800	872
Industrial coal-fired boilers	(average efficiency, %)	60	70	73	73[b]
Electric motors	(average efficiency, mean size, %)	87	90	92	92[c]

[a] Comprehensive consumption (not directly comparable with international statistics).

[b] Average U.K. efficiency for comparable size ranges.

[c] Average efficiency for U.S. high-efficiency models.

ciency: a) adopting modern and advanced processes and technologies, b) achieving economies of scale in production, and c) adopting modern management practices.

Advanced processes and technologies. There is much evidence that the adoption of new production methods and technologies contributed greatly to the modernization of Chinese industry during the 1980s and early 1990s while reducing energy use per unit of output. Concerted efforts are needed to ensure that this process of efficiency improvement continues into the future. In particular, policies should ensure that the most efficient new technologies and processes are brought to China, recognizing that at present, this may not be occurring in all cases. Table 2.5 lists improvements in industrial technologies and processes that can result in large reductions in energy intensity.

Economies of scale. Many, if not most, of China's major energy-consuming industries produce much of their output in suboptimal sized plants. Industries that can provide major energy savings in China through the adoption of larger plants include the following:

• *Cement*: The more than 5,000 small-sized plants should be gradually replaced and medium-sized plants should be phased out.

• *Electric power*: Old small- and medium-sized plants, based on low- or medium-pressure boilers, should be replaced by modern, larger, high-pressure units. Whereas state of the art 600 MW units consume about 310 gce/kWh (net), the small plants that still provide much of China's thermal power typically consume 500 gce/kWh or more.

• *Chemical fertilizer*: Small- and medium-sized ammonia plants, of which more than 1,000 make ammonium bicarbonate, should be replaced.

Modern management. To achieve the gains of new technologies and processes at optimal scale, Chinese enterprises must improve their management practices, including both "hardware" and "software." To achieve the energy-efficiency gains embodied in new processes, hardware is often required for automated monitoring, metering, calculation, adjustment, and control of new technologies. Equally important, yet often overlooked by enterprise managers, is the training of personnel to monitor, operate, and repair the new automated equipment. Research on Chinese industrial boilers has found that energy efficiency could be improved by 10–20 percent by optimizing the operation of existing boilers and ensuring that they are well maintained.

Table 2.5 Examples of energy savings through the adoption of new technologies and processes

Industry	New processes and technologies
Iron and steel	Conversion of open hearth furnaces to basic oxygen furnaces; use of DC electric arc furnaces; and phasing out of pig iron in steel making.
Non-ferrous metals	Replacement of outdated copper, lead, and zinc smelters; renovation of large facilities for electrolytic refining of aluminum; and adoption of modern technologies for new plants.
Cement	Replacement of wet kilns by dry process kilns with preheater or precalciner systems; phasing out of primitive shaft kilns; and utilization of industrial wastes such as fly ash and coal washery wastes.
Brick	Increase in percentage of hollow and perforated bricks, which reduces energy requirements for production and provides superior insulating properties for consumers.
Textiles	Greater application of electronic controls and other modern equipment.
Chemicals	Replacement of mercury by ion film process for caustic soda production.

Source: Energy Efficiency in China: Technical Analysis, report prepared for the China GHG Study, September 1994.

Industrial energy conservation projects. Projects that are considered "classic" industrial energy conservation projects typically focus on the addition, replacement, or renovation of specific equipment in the production process, with energy savings being the primary objective. The following five types of energy conservation projects have great potential in China:

• *Waste heat, gas, and resource recovery.* Waste heat can be used within the plant itself (e.g., cogeneration) or for district heating of nearby residential and commercial buildings. The recovered waste gas, large amounts of which are available from the iron and steel industry, can also be used by the plant or sold to gas utility companies. The recovery of other waste streams, from chemicals to fly ash, can be profitable and a major source of energy savings.

• *Cogeneration.* Many industries in China can realize large energy-efficiency gains by establishing cogeneration processes. Among the sectors that should make use of cogeneration are iron and steel, pulp and paper, and textiles.

• *Furnace and kiln renovation.* Many industrial processes require the use of furnaces and kilns; however, their efficiency is often quite low because of poor design, maintenance, or operation. In some cases, the kiln is the most important part of the plant, such as in the cement, brick, and glass industries. In such industries, improvement in the kiln requires major restructuring of the enterprise but also has the potential for major energy savings.

• *Energy management systems.* The installation of monitoring and control systems is often essential in reaping the energy-efficiency gains of new technologies. Opportunities for automation in China are numerous, including the textile industry and many industrial production lines. However, the training of personnel must accompany the installation of automated equipment for the efficiency gains to be realized.

• *Insulation, thermal/steam system renovation.* Significant improvements in energy efficiency can be realized through relatively simple and inexpensive measures, such as the insulation of pipes and the installation of effective steam traps.

Efficient equipment. Improvements in the energy efficiency of new models of widely used equipment can yield high energy savings for the economy. Small- and medium-scale coal-fired industrial boilers (which exclude boilers for power generation) account for

about one-third of China's coal use. Small- and medium-sized electric induction motors used in industry account for almost half of China's total electric power consumption. Improvements in the energy efficiency of such equipment can yield cost savings to enterprises, but these savings tend to be small in relation to overall costs. Nonetheless, even modest efficiency gains translate into enormous aggregate savings. Moreover, improvement in the efficiency of new equipment is important in the medium and long term for reducing greenhouse gas emissions. The efficiency of the equipment being installed now will be a key determinant of efficiency levels far into the future, especially given the current rapid growth in industrial capacity. This underscores the need to ensure that new installations utilize the most energy efficient technologies available.

Improvements in coal processing. Improving the quality of coal through cleaning, sorting, screening, and briquetting will increase the efficiency of coal utilization in China. These processes also offer the benefits of reducing coal transport costs, and, perhaps most importantly, lowering emissions of local pollutants such as TSP and SO_2. In 1990, less than 18 percent of all coal produced in China was washed. Coal washing can reduce both the ash and sulfur content of coal, and in the process allow coal-fired boilers to optimize the use of fuel and reduce the amount of TSP and SO_2 that is emitted during combustion.

Residential energy-efficiency improvements. An analysis of residential and commercial buildings in Shanxi, conducted as part of this study, shows that there is great potential for energy-efficiency gains in building construction through the use of hollow bricks, insulated plaster panels and walls, and improved glass and window seals.[10] District heating and other centralized heat distribution systems offer substantial improvements over traditional systems, such as small coal-fired stoves, when designed and operated properly. In addition to space heating, there is great potential for energy-efficiency improvements in residential cooking. While the use of gas, both piped and bottled, has increased in urban areas since the mid-1980s, the dominant fuel for cooking in urban areas is coal and in rural areas is coal and biomass fuels. The energy efficiency gained in switching from solid fuels to gas is largely because of the improved efficiency of the stoves and burners. Based

---

[10] See the project subreport prepared for this study, *Residential and Commercial Energy Efficiency Opportunities: Taiyuan Case Study*, September 1994.

on residential surveys in Shanxi, the average amount of energy used for cooking with raw coal and briquettes ranged from 437 to 579 kgce per capita, compared with LPG and coal gas, which used 69–114 kgce.

Under the Baseline GHG Scenario, rapid growth in energy use by the transportation sector over the coming decades, particularly for road vehicles, results in an increase in CO_2 emissions from about 4 percent of China's CO_2 emissions in 1990 to 5 percent by 2020. While still a relatively small share of China's total energy use, the quantity of energy consumed by China's transport sector will be huge in absolute terms and relative to transport energy use worldwide. Although no specific options for reducing transport energy consumption were evaluated for this study, technical efficiency improvements, modal shifts, and structural changes are all likely to be important for limiting the growth of GHG emissions from the transportation sector in China. Further research on specific options for GHG reduction in China's transportation sector is needed.

Alternative energy

China can greatly reduce greenhouse gas emissions by reducing the proportion of carbon-intensive energy sources in its energy mix. However, over the short term, there are limits to the extent of substitution for coal because of the long lead times needed to develop alternative technologies, the abundance of low-cost coal in China, and the sheer magnitude of the energy supply that will be needed to fuel China's economic expansion. Most experts expect that the only low-carbon-intensive fuels that can supply an appreciable increase in energy over the medium term in China are hydro for power generation, biomass for direct use or for power generation, and natural gas and coal-bed methane as direct substitutes for coal. Nonetheless, over the longer term, alternative energy is the only option for significantly reducing GHG emissions in China.

In the Baseline GHG Scenario, coal, petroleum, and gas provide 91 percent of China's energy supply and 78 percent of its electricity generation in 2020. Alternative energy supplies are used exclusively for electricity production. In addition to the Baseline GHG Scenario, the joint study team generated a high-substitution scenario to reflect the maximum amount of alternative (low-carbon) energy sources that could be developed in China by the year 2020 under current development trends. In the high-substitution sce-

nario, low-carbon fuels provide 39 percent of electricity generation, and an additional 130 mtce of low-carbon fuel is substituted for fossil fuels for non-power energy uses. The amount of coal used to produce electricity in the baseline energy demand scenario is about 1300 mt in 2020, accounting for approximately 30 percent of China's GHG emissions from energy consumption. Under the Baseline GHG Scenario, China will need to build an additional 700 GW of electric power capacity between 1990 and 2020, equivalent to the completion of 39 new 600 MW units each year. Nevertheless, non-power uses of energy, mainly for industrial process heat, residential cooking and heating, and transport, are estimated to account for more than two-thirds of primary energy consumption in 2020.

In the high-substitution scenario, alternative fuels including nuclear, wind, solar, and hydropower could provide nearly 40 percent of China's electricity by 2020, equivalent to about 16 percent of total energy. This would reduce carbon emissions by about 140 million tons (mtC) compared with the Baseline GHG Scenario. Direct substitution of fuelwood for coal could provide up to 75 mtce by 2020, which would reduce GHG emissions by about 55 mtC. Coal-bed methane could provide up to 40 mtce of energy by 2020, which would reduce GHG emissions by about 42 mtC.[11]

ELECTRIC POWER ALTERNATIVES

Hydropower. China currently has a large and expanding hydroelectric development program. Under the baseline alternative energy scenario, which assumes that 80 percent of China's hydropower resources are developed by 2050, hydropower capacity expands from 36 GW in 1990 to about 138 GW in the year 2020. Despite the large expansion in its capacity, hydro's contribution to power generation drops from 20 percent in 1990 to about 16 percent in 2020. Even assuming a program whereby China develops every economic hydro site by the middle of the 21st century (the high-substitution scenario), equivalent to about 184 GW of installed capacity by 2020, hydropower's contribution to total electricity production would still fall to approximately 19 percent by the year 2020.

---

[11] Emissions from burning 30 bcm of gas amount to 18 mtC, while the reduction in carbon-equivalent from using methane instead of flaring it amounts to 60 mtC. The difference—42 mtC—is the total GHG reductions from use of coal-bed methane.

Table 2.6 Electricity supply scenarios, 2020 (TWh)

	1990		Baseline		High-substitution	
	(TWh)	%	(TWh)	%	(TWh)	%
Hydro	126	20	601	16	719	19
Nuclear	0	0	208	5	568	15
Other renewables	0	0	45	1	208	5
Fossil fuels	495	80	2,996	78	2,355	61
Total	621	100	3,850	100	3,850	100

Nuclear power. Most Chinese energy experts emphasize nuclear and hydropower development as the principal alternatives to coal for future electric power generation in China. The baseline case assumes that China's nuclear capacity would grow from zero in 1990 to about 32 GW in 2020. Under the high-substitution scenario, China would have 87 GW of installed nuclear capacity by the year 2020, which would require China to complete more than ten 600 MW plants each year from 2010 onward. Cost issues aside, such an ambitious nuclear program would be the largest in the world and would require immediate action on technology development, personnel training, and the establishment of the necessary regulatory framework.

Other renewable energy. In the high-substitution scenario, the capacity of wind generators in China increases from the current level of 9 MW to 4,300 MW by 2010 and to 9,000 MW by the year 2020. At present, there is about 2,000 MW of wind-generated capacity worldwide. The high-substitution scenario assumes that as much as 82 GW of solar power generating capacity could be installed in China by the year 2020. In 1991, worldwide shipments of solar PVs were about 55 MW. Despite the great expansion of solar and wind generation under the high-substitution scenario, other renewable energy sources would provide only about 5 percent of electric power generation in 2020.

NON-POWER ALTERNATIVES

Fuelwood. While fuelwood plantations do not sequester much carbon on a net basis (Table 2.7), their contribution to CO_2 reduction can be significant if fuelwood is substituted for fossil fuels. Assuming that only about half of the 150 mtce of fuelwood currently consumed in China is produced on a sustainable basis, new sources of fuelwood can be used to reduce the overcutting and destruction of natural forests. Fuelwood from new plantations could amount to 276 mt (air dry), or approximately 150 mtce, by 2020. If half of this amount could replace coal, either for direct use or for power generation, GHG emissions could be reduced by about 55 mtC.

Coal-bed methane. Currently, only about 430 million cubic meters (cm) of methane, or less than 5 percent of methane emissions from large state-owned mines, is recovered through mine degasification and used. This amount could be increased to 2–4 billion cm (bcm) if the state mines with methane recovery systems could increase their recovery to levels of best-practice in China. If coal production in China expands to the levels envisioned by the Baseline GHG Scenario, the amount of coal-bed methane that could be recovered and used would be about 30 bcm, or double the current natural gas production in China.

Natural gas. Natural gas sources may be expanded either by discovery and development of new domestic sources, including tapping the large reserves of coal-bed methane, or imported by way of pipeline construction or shipments of liquified natural gas. It is assumed that in the future, natural gas will be used primarily for residential and commercial energy purposes, and, depending on the total supply, could also be used for electric power generation. The high value of natural gas in China warrants a more aggressive exploration and development program. Domestic natural gas production is assumed to rise from 15 billion cm in 1990 to 115 billion cm in 2020 under the baseline scenario, and to 150 billion cm under the high-substitution scenario. Both of these scenarios require large additions to proven natural gas reserves in China. Still, these amounts are equivalent to only 4 and 6 percent, respectively, of total projected energy use in China in the year 2020.

Other renewable energy. While small in comparison to overall energy use, other renewable energy sources can be important in specific applications, such as residential and commercial water and space heating, light industrial process heat, and water pumping in agriculture. Solar energy for residential and industrial water heating, for example, has significant potential for displacing coal and other carbon-intensive fuels for direct use.

Forestry

Through afforestation projects, the planting of timber and fuelwood plantations, and improved management of open forests, it is possible to store carbon in trees and soil and thus reduce net GHG emissions. Carbon sequestration is maximized by planting high-yield and fast-growing species on good land, under good growing conditions, and by applying modern forest management techniques. If some of the fuelwood can be substituted for coal, the reduction of CO_2 from the forestry sector in China would be even greater.

An analysis of tree planting in China (see pp. 50-53) shows that a large-scale afforestation program could

sequester 2.4–4.6 billion tons of carbon in woody biomass and soil over a thirty-year period, or an average of 80–153 mtC per year. Under the high scenario, the amount of carbon sequestered in the year 2020 would be 221 mtC—nearly 30 percent of China's 1990 GHG emissions and about 10 percent of baseline GHG emissions in 2020. China has planted trees at roughly the level of the "high scenario" since the late 1980s and has set a similar level of planting as the national goal for the year 2000. Nonetheless, even the low scenario requires an enormous program, including the planting and management of an additional 5 million hectares of forest land each year between now and the year 2020.

Agriculture

Through changes in agricultural practices, it is possible to reduce GHG emissions from the agricultural sector in China by 15–20 percent in the year 2020 compared with the Baseline GHG Scenario. Through the use of better cattle breeds and improved feed programs, it is possible to reduce methane emissions from ruminant animals. Rice cultivation techniques that limit the amount of time rice fields are flooded can also reduce methane production from rice fields.

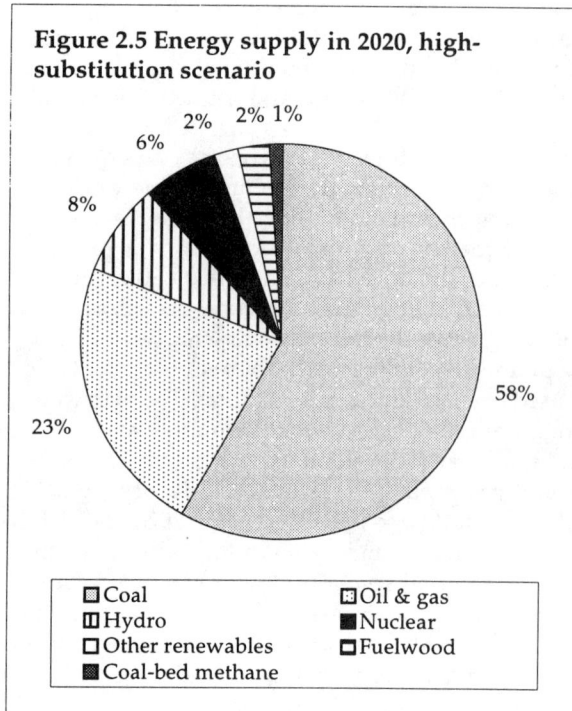

Figure 2.5 Energy supply in 2020, high-substitution scenario

- ☐ Coal
- ☐ Hydro
- ☐ Other renewables
- ▨ Coal-bed methane
- ☐ Oil & gas
- ■ Nuclear
- ☐ Fuelwood

Table 2.7 Cumulative carbon sequestration by plantations, 1990-2020 (million tons C)

Type of plantation	Scenario		
	Low	Medium	High
Intensively managed plantations	419	630	803
Extensively managed plantations	1,180	1,777	2,266
Fuelwood plantations	141	208	312
Open forest management	630	949	1,267
Total	2,370	3,564	4,648

Source: *Greenhouse Gas Emissions Control in the Forestry Sector*, subreport prepared for the China GHG Study, November 1994.

Improved ruminant production. Improving cattle by way of artificial insemination crossbreeding can increase the usable carcass weight, increase the fertility rate, reduce the number of bulls needed, and increase individual animal productivity. All of these improvements help to reduce CH_4 emissions per kilogram of product. There is the potential to reduce methane emission by almost 11 percent per animal. Chinese Ministry of Agriculture experts predict that improved breeding could extend to 70–90 percent of China's cattle population by the year 2020. If so, total methane emissions from ruminant animals could be reduced by about 10 percent compared with the Baseline GHG Scenario.

A second program, which shows great promise in China, is improving the productivity of cattle through the "ammoniated feed" program, whereby low-nutrient crop wastes are treated with ammonia and fed to cattle. The improved diet and increased feed consumption with ammoniated feed results in higher methane emissions from each animal, but emissions on a per kilogram weight gain basis decline by 35–40 percent. Therefore, fewer animals are needed to meet total food and animal power demands. Based on the current rate of adoption, it is possible that ammoniated feed could be used for about 20 percent of the cattle stock in China by 2020, which would reduce methane emissions from ruminants by about 15–20 percent compared with the Baseline GHG Scenario.

Changes in rice cultivation practices. A three-year study by the Nanjing Institute of Environmental Protection found that semi-dry rice cultivation can reduce CH_4 emissions by 31–43 percent. Because this system of cropping requires intensive soil preparation and water management, and can be implemented only on certain types of terrain, Chinese agricultural experts estimate that it could expand to only 15–25 percent of China's rice-growing area by 2020. If so, methane emissions from rice fields could be reduced by about 10 percent compared with the Baseline GHG Scenario. Field studies throughout the world have found that methane emission rates vary widely depending on field location and time of year. Chinese studies mirror this variability. Analyses of intermittent irrigation of paddy fields have exhibited methane emission levels 12 to 66 percent lower than for normal practices. Because many regions are too dry to employ this technique, the Ministry of Agriculture estimates that it could account for at most 10–15 percent of China's rice-growing area. This could result in a reduction in methane emissions from rice fields of about 8 percent compared with the baseline.

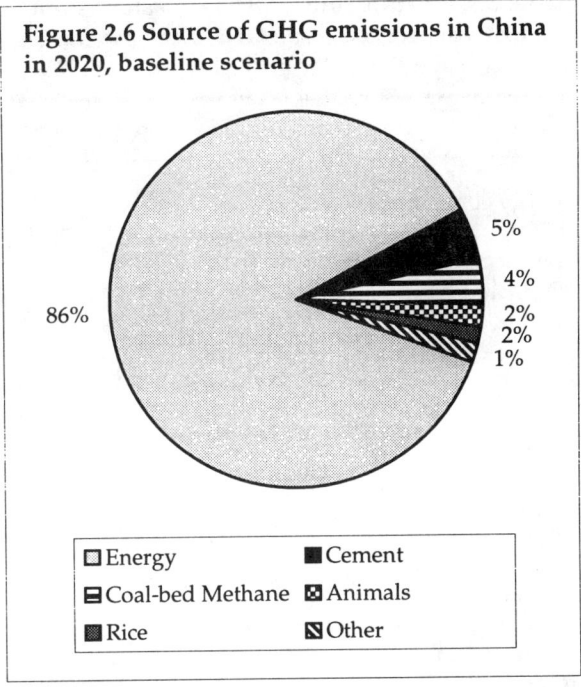

Figure 2.6 Source of GHG emissions in China in 2020, baseline scenario

86%
5%
4%
2%
2%
1%

☐ Energy   ■ Cement
⊟ Coal-bed Methane   ▨ Animals
▦ Rice   ◩ Other

CONCLUSIONS

From the macroeconomic analysis, it is possible to (i) estimate future GHG emissions in China and assess changes in the sources of emissions over time; (ii) examine the environmental implications of increased energy consumption; and (iii) identify the key options for reducing GHG emissions and assess the potential magnitude of emissions reduction.

Future emissions

Using the China GHG Model, the magnitude and composition of GHGs in China over the next twenty-five years can be analyzed. Under the Baseline GHG Scenario, the contribution of energy use to GHG emissions increases from 82 percent in 1990 to approximately 86 percent in 2020 (Figure 2.6). Given the rapid growth of the power sector under the Baseline GHG Scenario and the continuing importance of coal, emissions from the power sector increase from 24 percent to 31 percent of total CO_2 from energy use from 1990 to 2020 (Figure 2.7). The proportion of emissions from cement manufacturing, energy production (including coal-bed methane), and domestic animals remains fairly steady over time since all are correlated with the growth of the economy[12] (Table 2.8). Methane emissions from rice fields, which

Table 2.8 Possible future GHG emissions from China by source, Baseline GHG Scenario (mtC and equivalent)

	1990	2000	2010	2020
Energy consumption (CO_2)	650	987	1,512	2,045
Energy consumption (N_2O and CH_4)	14	21	33	44
Cement	29	63	86	108
Coal-bed methane	32	47	70	92
Animals	27	30	36	43
Rice	36	36	36	36
Other	20	24	29	37
Landfill gas	2	5	8	14
Fertilizer	2	3	4	4
Biomass burning (CH_4 and N_2O)	15	15	15	15
Oil and gas production (CH_4)	1	1	2	4
Afforestation	-7	-7	-7	-7
Total	801	1,201	1,795	2,398

Source: China GHG Model, joint study team.

are largely a function of the area of land planted rather than rice production, are not projected to increase substantially between 1990 and 2020. As a result, the share of GHG emissions from rice declines from about 5 percent in 1990 to 2 percent in 2020. Other sources of emissions remain at about 2 percent of total GHG emissions over the next twenty-five years. Landfill gas, which is correlated with the size of the urban population and the share of carbon-containing matter in urban wastes, is projected to increase sevenfold by 2020, but will remain at less than 0.6 percent of total GHG emissions. Despite the transport sector's rapid growth in energy use in the Baseline GHG Scenario, particularly for road vehicles, CO_2 emissions from that sector only rise from about 4 percent of total CO_2 emissions from energy use in 1990 to 5 percent by 2020. The amount of biomass burning, which is predominantly a function of residential fuel use in China, is not projected to increase in China over the coming decades, as rural households increasingly rely on other forms of energy for their needs.

---

[12] In the Baseline GHG Scenario, no measures are taken to reduce these emissions even though, for instance, it is likely that a growing share of the methane vented from coal mines will be collected and utilized.

Environmental implications of the Baseline GHG Scenario

Most of the assumptions used for the Baseline GHG Scenario, including the economic growth rate and the structure of final consumption, are derived from China's historical progress and the developmental experience of other Asian countries. Extraordinary improvements in energy efficiency, while unprecedented worldwide, are indeed possible based on the unique nature of energy consumption in China. However, the environmental implications of the Baseline GHG Scenario are staggering. Mining, transporting, and burning 3.1 billion tons of raw coal, roughly three times current usage, would have enormous consequences for air, water, and land quality.[13] Carbon dioxide emissions increase more than threefold in China between 1990 and 2020 under the Baseline GHG Scenario. Given this level of CO_2 emissions from China in 2020, the rest of the world would have to decrease current emissions by roughly a third to maintain worldwide emissions at 1990 levels.

Summary of options for reducing GHG emissions

What can be done to reduce such an increase in GHG emissions from China? The Baseline GHG Scenario already assumes enormous improvements in energy efficiency, and thus reductions in GHG emis-

---

[13] For example, the model shows that SO_2 emissions will increase from 16 mt in 1990 to 55 mt by 2020 while TSP emissions rise from 14 mt to 48 mt using 1990 emission coefficients (Table 2.9). Although structural change and environmental regulations will help control emissions, the analysis shows that great efforts will be needed to prevent the deterioration in air quality from increased coal use.

Table 2.9 Baseline GHG Scenario: Environmental implications of energy use

Year	1990	2000	2010	2020
Energy consumption (mtce), of which:	987	1,560	2,375	3,300
Coal consumption (mt)	1,053	1,574	2,376	3,100
TSP (mt)*	14	22	35	48
SO_2 (mt)*	16	27	41	55

* Assumes no change in emission coefficients from 1990.

Table 2.10 Potential for reducing GHG emissions in China, 2020

	GHG emissions relative to baseline (mtC)
Baseline GHG Scenario	2,398
High energy efficiency	-330
Alternative energy (high scenario)	-237
Afforestation (high scenario)	-221
Agriculture	-15 to -25

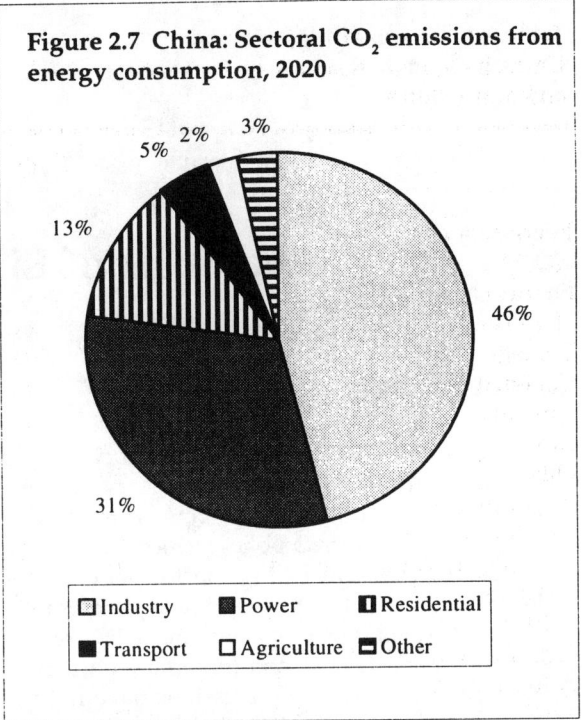

Figure 2.7 China: Sectoral CO$_2$ emissions from energy consumption, 2020

Industry · Power · Residential · Transport · Agriculture · Other

sions,[14] primarily as a result of economic structural change.

In addition to broad measures to improve the overall efficiency of the economy, there are more specific policy and investment options for reducing GHG emissions, including the following: (i) more aggressive technical energy conservation measures, (ii) more rapid adoption of low-carbon-intensive fuels, (iii) afforestation, and (iv) selected agricultural programs. The magnitude of reduction potential for these options, relative to the Baseline GHG Scenario, is given in Table 2.10. If all of these options were undertaken, GHG emissions could be reduced by more than 800 mtC by the year 2020—more than total GHG emissions from China in 1990. Instead of increasing more than threefold between 1990 and 2020, GHG emissions would increase by less than twofold. As will be discussed in the following chapter, however, the unit costs (yuan/ton) of undertaking each of these options vary greatly and the total costs for some are enormous.

---

[14] An approximation of the potential fall in energy intensity in China between 1990 and 2020 can be measured as the difference in energy use under the Baseline GHG Scenario (3,300 mtce) and the case where there is no change in China's I-O coefficients from 1990 (10,000 mtce).

Chapter 3

The Costs of Limiting Emissions

SUMMARY OF LEAST-COST OPTIONS

The most cost-effective options for reducing GHG emissions are those that already make sense in terms of financial, economic, or social returns to society; these are called *no-regrets* policies since they can be undertaken without incurring specific net costs for GHG emissions reduction. No-regrets projects usually provide several benefits, of which GHG reduction is only one. One reason that many no-regrets projects for reducing GHG emissions have not yet been undertaken, particularly in China and other developing countries, is because there are a variety of market and non-market barriers that restrain these investments.[1] In China, no-regrets projects for reducing GHG emissions include several energy-efficiency investments, forestry plantations, some projects to develop alternative sources of energy, and modifications to various agricultural practices. Depending on the level of GHG reductions desired, it may be necessary to undertake additional projects that have a positive net cost to China. Significant additional GHG reductions in China can be achieved through the further adoption of low-carbon-intensive energy sources, such as wind, nuclear, and solar energy. With current technology, however, large-scale adoption of these options is not part of the least-cost energy supply.

Costs of GHG reduction

Financial, economic, and, where possible, environmental economic analyses have been used to assess the cost of reducing GHG emissions.[2] The incremental cost of reducing GHG emissions should be net of

incremental benefits accruing to the enterprise and to Chinese society as a whole. Therefore, for purposes of calculating the net cost of reducing GHG emissions, financial, economic, and partial environmental benefits have been subtracted from total costs. The relative costs of various GHG reduction options in China are shown in Figure 3.1.

The net cost per unit GHG reduction is calculated by using standard cost-benefit analysis and calculating the GHG reductions associated with a project.[3] The resulting calculation is a net cost in yuan per ton of GHG reduced, expressed in present value terms. A major advantage of this approach is that the global benefits associated with reducing GHG emissions need not be quantified. This is an important consideration given the tremendous range of uncertainty connected with GHG emissions and global climate change. A second advantage of this methodology is that it allows comparison of the costs of very different types of projects, from an energy-efficiency investment in the steel industry to a forestry plantation for growing timber and sequestering carbon. Using the methodology outlined here, it should be noted that any GHG reduction project having a rate of return greater than the target rate of return will result in GHG reductions at no additional cost (zero or negative cost per ton of CO_2 reduced).

Significance testing

Projects must be deemed significant in terms of the level of GHG reductions that can be achieved. There may be a host of projects that are low-cost but which would have a minimal effect on GHG reduc-

---

[1] These issues are discussed later in this chapter and in Chapter 4.

[2] For details on the economic methodology used, see the project subreport prepared for this study, *Energy Efficiency in China: Case Studies and Economic Analysis*, December 1994.

[3] This is done by dividing the incremental net discounted cash flow by the present value of the incremental ton of GHG reductions at a given discount rate. This is referred to as "cost-effectiveness" analysis, since some of the costs (e.g., global damages from GHG emissions) are not quantified.

tions either because they are not replicable across the economy or because the GHG emissions related to the activity are insignificant relative to the economy's total emissions. The cut-off point for project significance is subjective; however, it should be high enough that scarce resources (capital, human, and managerial) are not used to develop projects that have little effect on total emissions. Estimates of reduction potential for the following options have been given in Chapter 2.

Energy efficiency

There is great potential in China for reducing energy use and CO_2 emissions at low cost through energy-efficiency investments. Twenty-five energy-efficiency projects, most of them in the industrial sec-

tor, were evaluated using case study analysis (see pp. 36-45). Using the results of the financial and economic analyses of these projects, the energy savings, GHG reductions, and net costs of GHG reduction were calculated. Since almost all of the energy-efficiency projects have high financial rates of return on a lifecycle basis, the net cost of GHG reduction is zero. However, there are several reasons why some of these projects are not sufficiently attractive to enterprises and investors.

Alternative energy

The use of low-carbon-intensive fuels, such as hydropower, wind, solar, nuclear, coal-bed methane, natural gas, and biomass energy, can yield large reductions in GHG emissions in China. However, many

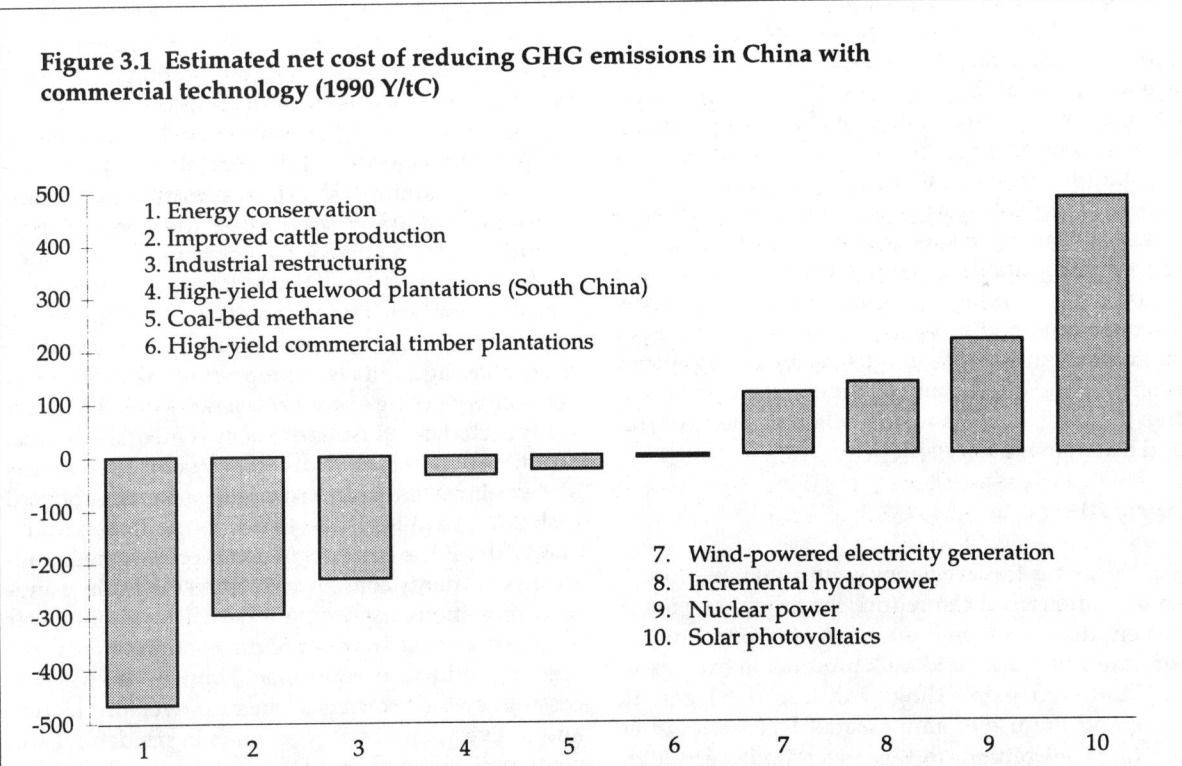

Figure 3.1 Estimated net cost of reducing GHG emissions in China with commercial technology (1990 Y/tC)

1. Energy conservation
2. Improved cattle production
3. Industrial restructuring
4. High-yield fuelwood plantations (South China)
5. Coal-bed methane
6. High-yield commercial timber plantations

7. Wind-powered electricity generation
8. Incremental hydropower
9. Nuclear power
10. Solar photovoltaics

Notes and sources: Net costs of reducing GHG emissions have been calculated using extended cost-benefit analysis. **1. Energy conservation:** Average cost per ton reduction for the seven classic energy conservation projects evaluated. See *Energy Efficiency in China: Case Studies and Economic Analysis*, December 1994. **2. Improved cattle production:** Average cost per ton for improved cattle breeding and feed programs. See *Greenhouse Gas Control in the Agricultural Sector*, September 1994. **3. Industrial restructuring:** Average cost per ton reduction for the six industrial modernization projects evaluated. Ibid. **4. High-yield fuelwood plantations:** See *Greenhouse Gas Emissions Control in the Forestry Sector*, November 1994. **5. Coal-bed methane:** Calculated from economic cost-benefit analysis data provided by the GEF China coal-bed methane project. **6. High-yield commercial timber plantations:** Average cost per ton of intensive and extensive timber plantations. See *Greenhouse Gas Emissions Control in the Forestry Sector*, November 1994. **7. Wind-powered electricity generation:** See *Alternative Energy Supply Options to Substitute for Carbon-Intensive Fuels*, December 1994. **8. Incremental hydropower:** Hydroelectric capacity beyond levels currently part of China's power expansion plan. Ibid. **9. Nuclear power:** Ibid. **10. Solar photovoltaics:** Ibid.

of these alternative energy sources have a positive net cost for reducing GHG emissions; that is, they are more expensive than coal or coal-fired power generation even when accounting for the local environmental costs of coal use, such as air pollution[4] (Figure 3.1). In the short to medium term, only hydropower, biomass fuels, natural gas, and coal-bed methane can be developed using conventional technology on a large scale at costs that are competitive with coal. Other alternative energy sources, including nuclear, wind, solar, and imported gas, are all projected to be more expensive than coal for power generation in China under typical conditions until at least the year 2020, even assuming substantial reductions in the cost of alternative energy.

Forestry

The analysis of tree planting for carbon sequestration in the United States and other developed countries has generally assumed that all costs (land acquisition, planting, and management) are attributed to carbon sequestration, and private, social, or other environmental benefits associated with the plantations are not calculated. The analysis here shows that planting trees on some types of plantations, particularly intensively-managed timber plantations, is profitable in parts of China on a life-cycle basis. For these types of commercial forestry projects, the net cost of carbon sequestration is zero. This result was obtained without considering the significant economic and environmental benefits associated with tree planting, such as wind shelter, erosion control, and ecosystem protection.

Agriculture

Increasing the efficiency with which ruminant animals are raised can reduce the amount of methane emitted. Economic analyses of two techniques for improving meat and milk production from cattle in China—cross-breeding with improved genetic stock and the use of ammoniated feed—show that the financial returns to these techniques are high. Economic analyses were not performed on semi-dry rice cultivation and intermittent irrigation of rice fields, which can also reduce methane emissions.

However, because these two practices are being used in some regions of China, it is assumed that the costs associated with reducing GHG emissions are low.

ENERGY EFFICIENCY

Introduction

This section reviews the economics of energy-efficiency investments in China, primarily in industry, and the incentives or disincentives for enterprises to pursue such investments. The analysis is based on background reports on energy efficiency for all major sectors, and on the results of twenty-five case studies of typical energy conservation investments. Both the background reports and the case studies were undertaken by joint Chinese-international teams.[5] The background reports describe the gains in energy efficiency that have been made in China over the past ten years and identify the factors within each sector that are most responsible for these gains.

The main objective of the case studies was to assess the economics of various energy conservation investments and, by subtracting the financial, economic, and local environmental benefits associated with these projects, determine the net cost of energy efficiency for GHG reduction. The case studies were selected to include projects with high potential for energy savings, but also to represent the range of investments yielding energy-efficiency gains. Each case study includes estimates of the incremental costs and benefits of a prospective or recently completed project at a specific Chinese enterprise, based on discounted cash flow analysis of life-cycle costs and benefits with and without the project. In addition to sixteen case studies of energy conservation investments in manufacturing enterprises, nine other case studies cover analysis of the economics of the production of more efficient industrial equipment, improved coal processing, energy conservation in power production and delivery, and energy savings in residential and commercial buildings.

---

[4] Alternative sources of power are compared to a new coal-fired power plant with high-efficiency particulate and NOx controls using low-sulfur coal. The inclusion of SO_2 scrubbers on new plants would most likely reduce the differential between coal and alternative energy sources; however, the cost advantage would still favor coal.

[5] Details of the technical and case study analyses are provided in three subreports prepared for this study: *Energy Efficiency in China: Technical and Sectoral Analysis*, August 1994; *Energy Demand in China: Overview Report*, February 1995 (forthcoming); and *Energy Efficiency in China: Case Studies and Economic Analysis*, December 1994.

CASE STUDY ANALYSIS: CONCEPTS AND METHODS

Financial and economic returns. Overall, the internal rates of return for the energy-efficiency investments in manufacturing enterprises are exceptionally high (Table 3.1). Based on the actual prices faced by Chinese enterprises in 1990, all sixteen projects reviewed have financial internal rates of return (FIRR) of well over 12 percent, while three-quarters of the projects have FIRRs of 20 percent or more. Over the life of the projects, therefore, financial benefits realized by the enterprises clearly far outweigh costs. In ten of the sixteen projects, the benefits include increases in output levels or quality of output, whereas in the other projects, benefits are derived solely from cost savings. Economic internal rates of return (EIRR) are also high. In these calculations, prices of inputs and outputs have been adjusted to reflect shadow prices, estimated to reflect the true values of these items in China's economy. The EIRR of all sixteen projects is 12 percent or higher.

The impact of price distortions. Although problems remain in certain sectors, the prices that most Chinese consumers pay for energy do not pose a major constraint to widespread adoption of energy conservation measures. By the end of 1993, 77 percent of all coal produced was sold at free-market prices, and price controls on the remaining 23 percent of production are scheduled to be removed in 1994. Although problems remain with the structure of electricity prices, the average electricity prices paid by consumers in most major load centers now approximate long-run marginal costs.

Table 3.1 Internal rates of return for selected energy efficiency projects in manufacturing

		Financial (%)	Economic (%)	Environmental economic (%)
M1.	Steel: Conversion of open hearth furnace to BOF	16	16	16
M2.	Steel: Adoption of continuous casting	19	19	20
M3.	Steel: Reheating furnace renovation	36	38	38
M4.	Steel: Blast furnace gas recovery	41	41	42
M5.	Aluminum kiln renovation	84	83	83
CH1.	Ammonia: Medium-sized plant restructuring	20	23	23
CH2.	Ammonia: Small plant waste heat recovery	71	54	56
CH3.	Caustic soda: Adopting membrane electrolyzer	29	32	33
B1.	Cement: Medium-sized kiln renovation	15	14	14
B2.	Cement: Conversion from wet to dry process	19	18	18
B3.	Cement: Small-scale kiln renovation	35	33	33
L1.	Pulp and paper: Adoption of cogeneration	25	29	29
L2.	Pulp and paper: Black liquor recovery	25	22	22
T1.	Textiles: Cogeneration in printing and dyeing	38	38	38
T2.	Textiles: Caustic soda recovery	58	34	35
T3.	Textiles: Computerized energy management system	infinite[a]	infinite	infinite

[a] The payback period is less than one year.

Source: Energy Efficiency in China: Case Studies and Economic Analysis, September 1994 subreport prepared for the China GHG Study.

Table 3.1 shows the effect that the use of economic shadow prices, as opposed to actual 1990 prices, has on project rates of return.[6] The returns of seven projects increase with adoption of prices more consistent with economic costs, while the returns of eight others decrease; one remains the same. However, the extent to which projects "switch" from being unattractive to attractive, or vice versa, is minimal. Since three-quarters of the projects reviewed exhibit rates of return of 20 percent or more under both the partially and fully reformed environments, it is clear that domestic price distortions alone are not the driving force leading Chinese enterprises to use energy-inefficient production processes. As a corollary, further price reform alone cannot be expected to resolve problems of inefficient energy use in industry.

Local environmental externalities. When reductions in health damages attributable to reductions in particulate and sulfur dioxide emissions are taken into account, the rates of return of the industrial projects reviewed (shown as "environmental economic" rates of return) increase in all cases over EIRR levels, although the average change is small, averaging 0.3 percent. For that reason, adding the domestic environmental benefits to the projects does not lead to a wholesale "switch" in any of the projects from unacceptable to acceptable because, in most cases, the rates of return of these manufacturing enterprise conservation investments are already high. Outside of manufacturing, however, the importance of local environmental benefits to the viability of energy-efficiency investments is greater in some cases, most notably in coal processing.

INDUSTRIAL ENERGY EFFICIENCY

Three types of industrial energy-efficiency projects were analyzed in the study, and most of the case studies reviewed below fall into one of the following categories of projects.

- Industrial restructuring;
- Industrial energy conservation projects; and
- High-efficiency energy-consuming equipment.

Coal processing, which can improve energy efficiency in industry as well as in other sectors, is addressed in a separate section.

---

[6] Energy price distortions were greater in 1990 than in 1993.

Industrial restructuring projects

Many of the projects that yield high energy savings in manufacturing require major restructuring of existing plants. Such programs are common throughout China today, as enterprises seek to upgrade production processes and capture economies of scale to better compete in the emerging market economy. Among the case studies, six are typical of industrial restructuring projects, including (i) conversion of open-hearth steel furnaces to basic oxygen furnaces (BOF), (ii) adoption of continuous casting in steel production, (iii) process and product restructuring in the medium-scale ammonia industry, (iv) large-scale kiln renovations in medium-sized cement plants, (v) conversion from the wet to the dry process in cement production, and (vi) adoption of membrane electrolyzer technology in caustic soda production.

Project characteristics. As shown in Table 3.2, industrial restructuring projects with high energy-efficiency gains typically exhibit the following characteristics:

Relatively modest rates of return. Of the sixteen case studies in manufacturing, all four projects with financial rates of return under 20 percent are restructuring projects.

High investment costs. Since these projects require major retooling of production lines, investment costs typically exceed Y100 million (US$21 million).

Long payback periods. Because of modest rates of return and high investment costs, discounted payback periods of the projects reviewed are six to nine years.

Multiple benefits. Restructuring projects generally provide a variety of benefits, including reductions in operating costs, of which energy costs are just one. Most importantly, such projects generally yield increases in output value because of increased production levels, quality improvements, or both. Output value increases substantially in all six of the restructuring case studies.

High financial significance to enterprises. Net project benefits are high relative to the enterprise's overall operations: they are equivalent to at least 20 percent of total enterprise cost streams over the project period. Indeed, successful completion of these projects is often critical to the enterprise's overall financial health.

Table 3.2 Economic parameters of industrial restructuring projects with energy-efficiency gains

	FIRR (%)	Discounted payback period (years)	NPV of investment (million yuan)	Output value increase (%)	Significance ratio[a] (%)	Annual energy savings ('000 tce)	Energy savings ('000 Y/tce)	Financing burden[b]
Steel: BOF conversion	16	9	741	15	2[c]	145.6	5.09	29
Steel: Continuous casting	19	8	298	125	4[c]	139.3	2.14	87
Ammonia: Medium-sized plant restructuring	20	8	298	125	44	139.3	2.14	87
Cement: Medium-sized kiln renovation	15	9	362	128	47	93.5	3.87	122
Cement: Conversion to dry process	19	9	362	128	47	93.5	3.05	75
Caustic soda: Restructuring	29	6	44	20	22	17.6	2.5	124

[a]This ratio measures the financial significance of a project's net benefits to the enterprise. It is defined as the NPV of the incremental cash flow of the project divided by the PV of enterprise total cost stream without the project during the time period that the project would be undertaken.

[b]This is a rough measure of the capacity of the enterprise to finance the project. It is the percentage of the enterprise net cash flow without the project, over the same time horizon as the proposed project, required to finance the project investment.

[c]These case studies yielded low significance ratios because the projects considered only one steel shop in large companies. The first case involves the Anshan Iron and Steel company, the largest in China, and the second case was at the Benxi Iron and Steel Works, also a large plant.

High energy savings per project. Because industrial restructuring projects tend to be large, energy savings in the case studies reviewed average 185,000 tons of coal equivalent (tce) per year.

High gross investment costs per unit energy savings. Because energy savings are just one of several project objectives, the gross investment cost per tce saved tends to be high, averaging more than Y3,000 in 1990 prices.

High financing burden. Enterprise financing of industrial restructuring projects is problematic, not only because investment costs are high but also because, prior to restructuring, these enterprises tend to be financially weak.

INCENTIVES AND CONSTRAINTS TO IMPLEMENTING RESTRUCTURING PROJECTS

There is now great incentive for enterprises in China to engage in restructuring projects. Often, they represent the cornerstone of enterprise management's plan to improve its competitive standing. Indeed, completion of restructuring projects may be required for enterprise survival. In many cases, such as the steel, cement, and ammonia projects reviewed, projects are not especially risky from a technical viewpoint. Typically, other enterprises have already pursued similar technical renovations and their experience is fairly well known.

Securing project financing is probably the most

serious constraint in implementing these restructuring projects. Upfront investment costs are high, and investment capital is in short supply. Moreover, many of these enterprises are either currently incurring losses or are only marginally profitable—hence, the need to engage in a restructuring of operations. Moreover, because significant financial resources are required for restructuring, internally generated enterprise investment funds are often insufficient. Because enterprises are often saddled with high fixed operating costs stemming from a range of social guarantees to workers, the financial risk associated with these projects is high. The success of industrial restructuring efforts ultimately depends on the current enterprise reform drive, and especially on the continued reform of China's banking system.

Industrial energy conservation projects

Seven projects reviewed as case studies fall into the category of energy conservation: (i) renovation of reheating furnaces in the steel industry, (ii) recovery of blast furnace gas for use in cogeneration in iron and steel production, (iii) waste heat recovery in small-scale ammonia plants, (iv) renovation of small-scale vertical shaft cement kilns, (v) cogeneration in pulp and paper manufacturing, (vi) cogeneration in textile printing and dyeing operations, and (vii) adoption of computerized energy management systems in textile plants.

Project characteristics. "Classic" industrial energy conservation projects typically exhibit the following characteristics (Table 3.3):

High rates of return. The financial rate of return of each of the seven projects reviewed is at least 25 percent.

Modest investment costs. These projects are smaller than restructuring projects. Investment costs in the cases reviewed averaged Y12 million (US$2.5 million in 1990 prices).

Table 3.3 Review of economic parameters of industrial energy conservation projects[a]

	FIRR (%)	Discounted payback period (years)	NPV of investment (million yuan)	Output value increase (%)	Significance ratio[b] (%)	Annual energy savings ('000 tce)	Investment per tce saved ('000 Y/tce)
Steel: Reheating furnace renovation	36	4	0.89	0	...	1.45	0.61
Steel: Blast-furnace gas recovery	41	6	14.11	0	3.2	24.91	0.57
Ammonia: Small plant waste heat recovery	71	3	2.69	11	6.2	4.72	0.57
Cement: Small-scale kiln renovation	35	6	23.63	32	12.2	40.33	0.59
Pulp and paper cogeneration	25	7	36.62	0	...	33.48	1.08
Textiles: Cogeneration	38	4	6.64	0	8.6	3.63	1.83
Textile: Energy management system	infinite	0.5	0.11	0	1.5	0.42	10.50

[a]The aluminum kiln renovation (M5), pulp and paper black liquor recovery (L2), and textile caustic soda recovery (T2) case studies are not listed in Tables 3.3 or 3.4, since these projects exhibit mixed characteristics. They are classified between the restructuring and "narrower" energy conservation categories.

[b]See Table 3.2 for explanations.

Medium-length payback periods. Although there is substantial variation, discounted payback periods are shorter than for restructuring projects, averaging four to five years in the cases reviewed.

Focus on energy benefits. In keeping with their narrower focus, the primary financial benefit of energy conservation projects is a reduction in energy operating costs. Although these benefits alone generate high rates of return in the cases reviewed, most projects exhibit no increase in production levels or value of commodities produced.

Low financial significance to enterprises. Net project benefits tend to be low in relation to the enterprise's overall operations since these are generally small projects. With the exception of pulp and paper cogeneration and small-scale cement kiln renovation, the net project benefits of the projects listed in Table 3.3 are equivalent to less than 10 percent of total enterprise cost streams over the project period.

Modest energy savings per project. Energy savings in the case studies reviewed average just over 15,000 tce per year. In the aggregate, however, implementation of many industrial energy conservation projects may lead to greater energy savings than in restructuring projects because of widespread application.

Low gross investment costs per unit energy savings. The gross investment cost per tce saved tends to be substantially lower than in restructuring projects, averaging less than Y800 (1990 prices) per tce saved per year.

Modest financing burden. Enterprise financing of these conservation projects tends to be easier than for restructuring projects. In the cases reviewed, investment costs were typically 20 percent or less of the net cash flow of the enterprise without the project.

ENERGY CONSERVATION PROJECTS IN UNHEALTHY ENTERPRISES

Because "classic" industrial energy conservation projects focus on equipment replacement or renovation of just one or several aspects of a plant's operation, they may be described as "incremental" projects, rather than restructuring projects. Even though these incremental projects show good financial and economic returns, they may exist in enterprises that are financially and economically nonviable. Indeed, the greatest potential for high rate-of-return, efficiency-oriented incremental projects tends to be in inefficient

enterprises. The existence of high-return, yet uncompleted, incremental projects may be one symptom of broader, more fundamental issues requiring the completion of broader restructuring efforts, or, at worst, plant closure.

Industrial energy conservation projects, therefore, must be evaluated in the context of the overall operations of enterprises. Two issues to consider when reviewing these incremental projects are whether the enterprise will be viable after the project is undertaken, and whether the enterprise can finance and manage the project. Isolated support for incremental projects in unhealthy enterprises is both counterproductive and a waste of resources. Energy conservation projects may, however, be a productive part of a broader restructuring package to make an unhealthy enterprise viable, especially in energy-intensive industries.

INCENTIVES AND CONSTRAINTS TO CONSERVATION PROJECT IMPLEMENTATION

As mentioned above, some seemingly attractive energy conservation projects may not be undertaken because of fundamental financial and economic problems in an enterprise. In addition, however, there are numerous constraints impeding the implementation of attractive energy conservation projects in enterprises that are financially healthy and growing. Some of these constraints stem from the current transition from a planned to a market economy, while others are common in developed market economies as well.

Weak cost-consciousness. The "classic" energy conservation projects presented in Table 3.3 depend on upfront investments to reduce future operating costs. Such investments require a commitment by enterprise management to control operating costs as a means to increase enterprise profits. While the profit motive and market competition are rapidly increasing cost-consciousness in Chinese enterprises, it takes time for management methods and attitudes to change.

Reforms are underway in China to provide state-owned enterprises with autonomy and full accountability for profits and losses. New fiscal and tax provisions will enforce budget constraints and accelerate the move toward corporate forms of operation. Implementation of these reforms takes time, however, as enterprise managers adapt to the new competitive environment. Thus, the provision of well-designed and practical information on the cost-savings and profit implications of specific energy conservation investments is both urgent and important.

Demands for short payback periods. Enterprise managers and potential investors almost always assess "classic" energy conservation projects in terms of their payback period, rather than their life-cycle rate of return. Conservation investments with payback periods of more than five years are rarely undertaken by enterprises or investors, unless there are other pressures to do so.

Risk and uncertainty. The use of payback period calculations is usually associated with risk and uncertainty. Whereas investments in output expansion yield physical, easily perceived benefits, the benefits from energy-saving investments are future cost savings calculated by analysts. Skepticism concerning these calculations often stems from perceptions of technical risk. Will the renovation project or new equipment really result in the energy savings forecast? In China, this natural bias has been exacerbated by a management culture attuned to maximizing physical output and achieving—or surpassing—output quotas.

Additional uncertainties or risks may be seen on the other part of the cost savings calculation: forecast energy prices. Although recently energy prices have increased in real terms, the pricing system has been changing rapidly. Other aspects of the system, especially for electricity, also remain complex, non-transparent, and difficult for enterprise managers to fully understand and predict. If non-energy prices (such as prices for a factory's products) increase faster than energy prices, the original energy conservation investment may be less attractive than other uses of investment funds.

A final uncertainty is market and technical risk stemming from the rapid change in China's economy. Increasing competition and market development are driving a major economic restructuring and causing volatility in the profitability of different sectors and product lines. In China today, flexibility to respond to rapid market changes is critical. Long payback periods mean funds are tied up. Not only may future opportunities for high-return profits be lost, but market changes may make renovated production lines, although more energy-efficient, less profitable or unprofitable as a whole.

Inadequate information. Often, viable energy conservation investments are not undertaken because enterprise managers have limited knowledge of these opportunities. Particularly lacking is practical information on enterprise experiences with different technologies and the details of cost savings and increased profits.

Low financial significance and transaction costs. Even if rates of return are high and payback periods are considered acceptable, many energy conservation projects may not be considered priority investments by enterprises because net benefits are not large relative to the size of the enterprise. The time and effort required for staff to gather information, design projects, undertake the relevant analysis, and implement the projects may not be worthwhile. In other words, there may be additional, hidden costs involved in project preparation and implementation, especially for highly valued staff, that reduce the project's attractiveness.

High-efficiency energy-consuming equipment

The introduction and dissemination of high-efficiency equipment will entail technology transfer, product development, and manufacturing and marketing by producers, as well as more purchases by consuming enterprises. The perspectives of producers and consumers and the economic issues they face are briefly reviewed below.

EQUIPMENT PRODUCERS

The equipment industry in China includes a few large key producers and many small producers, most of which lack economies of scale and use outdated technology. Product prices are now largely decontrolled and a competitive market framework is in place. However, industries are still adjusting to market forces. These industries have not completely realigned their structure, supply and demand remain unbalanced in some cases, and future market price levels are uncertain. Product marketing capabilities and customer service efforts are still weak, since they did not exist under the planned economy. Some key issues relating to the financial attractiveness of producing more energy-efficient equipment models, based on the case study analysis, are presented below.

Market conditions and enterprise profitability. Production capacity exceeds market demand in the industrial boiler industry, profits on the standard models are weak, and enterprises are exploring new opportunities. However, demand for variable-speed electric motors and steam traps exceeded supply in early 1993. In two case studies, the rates of return on investments to expand variable-speed motor production were very high under prevailing product prices. The relationship between supply and demand for new, high-efficiency motors is unclear, and the returns to investment for expanded production at current prices appear modest.

Product pricing. Prices for more efficient products should divide benefits between producers and consumers in such a way that each has a proper incentive to pursue the innovation. This is best achieved by giving full play to market forces. Exact production costs for a new generation of more efficient industrial boilers have yet to be determined, but to be commercially successful, product prices cannot be significantly higher than existing designs. For more efficient variable-speed motors, market-clearing prices have not been fully realized.

Project rate of return sensitivity. The case studies show that the rates of return for investment in expanded production of high-efficiency and variable speed motors are particularly sensitive to product prices and raw material costs. The relative attractiveness of developing new product lines must be compared with upgrading existing ones.

Foreign participation. Several joint-venture operations have been established in the electric motors industry, where export potential is strong and Chinese production costs are low. In the coal-fired industrial boiler industry, however, there are no joint ventures with foreign firms. There is no market for these boilers abroad, and intellectual property rights would be difficult to protect in the domestic market.

CONSUMING ENTERPRISES

Investments in higher-efficiency equipment generally exhibit the following characteristics:

Attractiveness varies widely in different applications. The ratio of operating costs (primarily energy costs) to equipment purchase costs is high for higher-efficiency equipment.[7] The greater the total energy costs, the greater the benefit for a given percentage increase in efficiency. Project attractiveness is sensitive to the number of operating hours, load rates, and the level of energy prices. In cases where pollution already translates into financial costs, the attractiveness of more efficient industrial boilers varies according to environmental protection requirements.

Attractiveness varies greatly between new installations and replacements. The attractiveness of purchasing new high-efficiency equipment is greater when a choice is being made for a new installation. In replacing older operating equipment, the key parameters are the degree of relative efficiency improvement and the remaining value (e.g., operating life) of the equipment being replaced.

Low financial significance. Although rates of return may be high and payback periods short, the financial significance to enterprises of the reduced energy costs from investments in higher-efficiency equipment is typically quite low. This is particularly true for items such as small electric motors or steam traps. Except for adoption of variable speed motors in certain applications, consuming enterprises may often consider energy efficiency less important than other product characteristics, such as reliability, product life, ease of installation, and ease of purchase and repair.

CONSTRAINTS TO DISSEMINATION OF HIGH-EFFICIENCY EQUIPMENT

Producing enterprises. Barriers to technology transfer from abroad represent one set of constraints to expanded development and production of more energy-efficient equipment. In all of the equipment industries, application of foreign technology is critical in achieving high efficiency at reasonable cost. Although necessary to enable proper compensation for research and development investment, protection of intellectual property rights presents a number of barriers. First, foreign firms are often concerned that protection will be inadequate, and thus deny access to state-of-the-art technology. Second, the costs of such rights may often be prohibitive to individual Chinese enterprises, especially considering the additional risk associated with cost recovery through higher prices. Finally, because by definition successful protection of intellectual property rights means that technology access is limited to only some producers, the overall objective from the perspective of energy efficiency and greenhouse gas reduction—widespread and rapid dissemination—may be compromised.

---

[7] This ratio also is substantially higher in China than in most other countries, since prices for the equipment are substantially lower than prices abroad (although equipment quality and reliability are generally poorer).

Another category of constraints is the risk and uncertainty involved in developing and marketing the new products. Some risk and uncertainty is to be expected as a normal aspect of business development. However, uncertainties concerning product cost and market response at different prices were especially high in some cases reviewed, such as for high-efficiency coal-fired industrial boilers or high-efficiency electric motors. For boilers, where new models must be developed specifically for Chinese conditions, major technical risks exist in product development and initial manufacturing. Production costs for much of the new equipment are uncertain, including the future price and availability of raw materials. Generally speaking, however, the largest risk concerns marketability and future product price levels, especially because competitive markets for producer goods are relatively new and consumer interest is uncertain for the reasons given below.

Consuming enterprises. Although there are cases where payback periods for installation of new high-efficiency equipment are quite short and consumer interest may be strong, consumer adoption is often bound by the same types of constraints as are classic industrial energy conservation projects: enterprises may not be sufficiently cost-conscious, investments may be perceived as too risky, or enterprises may not have good information. Even more so than for other conservation projects, incentives for the installation of high-efficiency equipment are weak because of low financial significance. Many enterprises will give low priority to such investments because the potential cost savings are low relative to total costs. Moreover, particularly for motors and associated electrical equipment, new models may not conform to existing factory layouts, making enterprise adoption technically difficult.

Improved coal processing

The economics of expanded washing of steam coal depend upon coal varieties, transportation distances between producers and consumers, how coal is used, and the nature of air pollution concerns at consumption sites. Transportation cost savings, stemming from the need to move fewer tons of coal per unit energy because of the lower ash content in washed coal, are substantial. Additional, but smaller, benefits include some energy-efficiency gains, and reduced maintenance and ash disposal cost savings for consumers. Substantial benefits may be achieved in environmental protection, in terms of improved particulate and SO_2 control. This is especially so in the case of small boilers, which may often have poor, if any, emissions control equipment. One recent study shows that a major increase in steam coal washing is economically justified based on transportation, user efficiency, and maintenance gains alone. More thorough cleaning is justified if environmental benefits are included.[8]

From the producer's perspective, the attractiveness of steam coal washing varies dramatically according to sales price. Two washery projects analyzed for this study showed a wide spectrum, yielding an unviable financial rate of return in one example and an exceptionally high rate of return in the other. From the consumer's perspective, there is little interest in paying significantly more for washed steam coal.

In the case of briquettes, the manufacture and use of household briquettes is generally attractive financially, whereas the manufacture and use of industrial briquettes, based on existing pilot applications, is generally not attractive unless environmental benefits are considered. The use of relatively simple briquettes in specially designed, low-cost *household briquette* stoves is common in urban areas, and further extension in rural areas will provide consumers substantial benefits in terms of improved efficiency, convenience of use, and less indoor smoke. Manufacturing techniques for *industrial briquettes* or pellets have been less successful, however, and efficiency and boiler maintenance gains have not been sufficient to justify the higher consumer prices. Nonetheless, the potential for environmental benefits, especially improvements in health, is considerable from lower particulate and SO_2 emissions.

IMPLEMENTATION ISSUES AND CONSTRAINTS

Achievement of efficiency and environmental protection gains through improved coal sorting, screening, and allocation, and through increased beneficiation (such as washing and briquetting) must involve the entire supply-use chain: coal producers, coal transporters, wholesale and distributing agencies and companies, and final consumers. However, the existing coal allocation and marketing system, including pricing, in China does not provide proper incentives to improve coal quality. Substantial progress will require a strong and sustained commitment to (i) further reform markets for coal supply

---

[8] See *China: Investment Strategies for the Coal and Electricity Delivery Systems* (Washington, DC: World Bank), April 1994.

and distribution, and (ii) improve the regulation of air pollution emissions, especially enforcement.

A shift from the traditional supply-driven system to a consumer demand-driven system is critical to ensure that producers and consumers receive the proper signals concerning the most efficient mix of coal products. Until recently, China's coal production and allocation system was characterized by command and control for production, central allocation of mine output and transportation, and consumer acceptance of whatever coal was supplied. In 1993 and 1994, prices in the coal industry were deregulated, a significant step toward adopting a more market-oriented system. Yet, major institutional changes and progress in enterprise reform throughout the supply and distribution chain will be necessary for the new system to work. Consumer attitudes will need to change, and enterprises accustomed to a "take-it-or-leave-it" system will have to get used to the idea of choice and discrimination between different products and different prices.

For an efficient market-based system to work, there must be standard, well-recognized procedures for preparation and enforcement of long-term contracts between consumers and suppliers. Long-term assurances for supply of a given type and quality of coal are critical for large and medium-sized consumers to maximize the efficiency of coal combustion. For China's coal market to develop, there will be a need for more sophisticated institutions, expertise, and information dissemination mechanisms relating to a wide variety of coal types and products with ever-changing product-specific prices.

The growth of a market-oriented system, however, will not by itself provide an adequate framework for the development of environmentally sound coal supply and use patterns. The experience of other countries has shown that improvements in coal processing, such as the washing of steam coal, are achieved only when strong emissions control regulations are in place. Environmental regulation is the key to changing consumer demand, which in turn changes the coal supply and distribution chain. In recent years, China has made considerable progress in establishing air quality standards and emission regulations. These regulations, however, need to be further improved and broadened in scope. Most importantly, enforcement must be improved. Significant improvements in emissions control can be expected only when polluting enterprises face fines and other financial incentives to reduce their pollution.

Energy-efficiency improvements in buildings and the residential sector

Improvements in the construction and maintenance of residential and commercial buildings are closely related to China's system of housing ownership. In urban areas, below-market rents for housing and subsidized utility bills have typically been provided to workers by government work units and enterprises as part of their compensation. Housing and wage reform are necessary to improve the maintenance and reduce utility expenses of residential buildings.

A second problem, which also exists in well-developed market economies, relates to building construction. When new housing is built, contractors and builders often try to minimize construction costs, including investments in more energy-efficient materials and systems, because they will not be responsible for the long-term operating costs. In China, this problem is compounded by the rapid increase in housing construction. Contractors want to build as quickly as possible, and incentives for innovation are weak since the demand for buildings exceeds the supply. Building material producers also have little incentive to create new products, since the demand for old products is high. A case study of residential building construction in Beijing using new building standards shows that the project is viable with an internal rate of return of 25 percent based on projected energy savings relative to investment cost. Yet, achievement of such gains on a large scale will require major efforts to further develop market forces and regulation to increase the incentives for all of the parties involved.

ALTERNATIVE ENERGY

In the short to medium term, until 2010, it will be difficult for China to reduce its reliance on coal by adopting less carbon-intensive fuels. Over the longer term (beyond 2010), alternative energy sources could be developed in China on a large scale, providing some 20 percent of total energy and 40 percent of total electricity supply by 2020. However, the analysis conducted for this study shows that even under optimistic assumptions, most of the alternative energy sources that can be developed on a large scale in China over the next twenty-five years are more costly than coal.[9] Unless the costs of alternatives can fall to a level comparable to coal, there will be a high net cost for reducing CO_2 through the adoption of alternative energy.

The cost analysis summarized in this section and the GHG reduction potential estimates provided in Chapter 2 are the product of a joint international-Chinese research effort. International experts prepared reports on the current development of liquified natural gas (LNG), solar photovoltaic (PV), nuclear, and wind energy, focusing on recent commercial developments and the current and projected future costs of these technologies internationally. Chinese experts evaluated the current stage of development of these and other alternative energy technologies in China and prepared scenarios of future energy supply and costs for the years 2000, 2010, 2020, and 2050.[10]

The environmental analysis of alternative energy sources in this section requires some elaboration. First, because coal is the dominanat source of energy in China, it has been used as the reference for comparing the costs of alternative energy sources. A premium has been added to the market price of Chinese coal and the cost of coal-fired power generation to allow for the removal of particulates and waste gases to meet international standards. Second, negative environmental impacts of alternative energy sources have not been calculated. For instance, while most alternatives to coal that reduce CO_2 emissions also reduce some local air pollutants (such as TSP, SO_2, and NOx), there can be harmful environmental impacts associated with alternative fuels. Before adopting alternative energy technologies, major environmental costs should be assessed, such as resettlement and ecosystem damage associated with construction of hydroelectric projects; the hazards of LNG transport and distribution; local air pollution associated with biomass combustion; and the costs and risks of securing, storing, processing, or disposing of nuclear fuel.

Prospects for alternative energy technologies

Short to medium term. The mix and quantity of alternative energy resources that could be developed on a large scale before 2010 are limited to those that are already available in China or could be easily imported from abroad. Coal is projected to be the domi-

nant fuel for electric power generation and direct fuel use in China. The joint study team's high coal-replacement scenario projects that only 29 percent of the total electricity generated in 2010 could be obtained from sources other than coal. There are two main reasons why alternatives for power generation over the short to medium term will be limited. First, substantial lead times are needed for capital-intensive projects such as hydro and nuclear power. Second, some alternatives may not be fully commercial before 2010. Of the non-coal-fired power supply in 2010, 80 percent is predicted to come from hydroelectric generation and the remainder from all other sources. The only alternatives to coal that could provide significant amounts of energy for direct use before 2010 are coal-bed methane, biomass, natural gas, and, possibly, solar technology.

Longer term. While alternative fuels could provide up to 40 percent of China's electricity supply by 2020 (as presented in the high-substitution scenario), substitutes to coal for power generation will be costly given current expectations of future technology development. The joint study team estimates that, compared with the baseline scenario, it will require an additional 750 billion yuan (US$159 billion) to meet electric power demand in the year 2020 under the high alternative energy scenario.[11] While there are low-carbon substitutes for coal in electric power generation, such as hydro, nuclear, wind, and solar, more than 80 percent of the energy used in China is directly consumed for process heat or for residential cooking and heating. Even with the accelerated growth in electric power sector in China, direct use of energy will still account for about 60 percent of total commercial energy in China in the year 2020.

Cost estimates for low-carbon energy sources

Based on a review of international and domestic experience, the joint study team has estimated the costs of alternative energy sources. For electric power generation, where alternative energy options are greatest, both investment and levelized generation costs have been estimated. Because large-scale substitution will not take place before 2010, the time frame for the cost estimates is roughly the year 2020. Ranges have been used to reflect the high degree of uncertainty for most of these estimates. Cost esti-

---

[9] The cost of coal use in China is assumed to increase, both for direct use and for electric power generation, as a result of more strict environmental controls on coal combustion.

[10] A complete review of alternative energy in China and its role in reducing GHG emissions is contained in the project subreport prepared for this study, *Alternative Energy Supply Options to Substitute for Carbon-Intensive Fuels*, December 1994.

---

[11] All cost estimates in this chapter are in 1990 constant prices and converted to US$ at the 1990 official exchange rate of 4.7 yuan/US$.

mates for alternative energy sources that could substitute for coal or other fossil fuels for direct use have been based on the economic analyses conducted for this study and other GEF projects.[12]

COAL SUBSTITUTES FOR ELECTRIC POWER GENERATION

Under all alternative energy scenarios, the absolute amount of thermal power capacity in China will increase over the next twenty-five years. The extent to which low carbon-intensive fuels can be substituted for coal for power generation is related to upfront capital investment and capacity cost, and to the average cost of generating electricity for each of the alternatives.

Coal as the reference. Alternative energy sources must be competitive with coal. According to Chinese estimates, investment costs for domestic coal-fired plants are about 2,600 yuan/kW (1990 yuan), while an additional 15 percent will be required to meet stricter environmental regulations; total investment is thus estimated at about 3,000 yuan/kW. Costs for coal-fired power plants in China with some foreign equipment and advanced environmental controls are approximately 4,500 yuan/kW.[13] Based on these capital investment costs and long-run costs for coal, the levelized generation costs for a coal-fired plant in China have been estimated in the range of 0.18–0.25 yuan/kWh (1990 constant prices) in 2020.[14]

Hydro. Hydropower is currently part of China's least-cost development program for electric power generation. Under the Baseline GHG Scenario, which is an extrapolation of China's current development program, China will quadruple its hydroelectric generating capacity by the year 2020. If hydro resources are developed even more rapidly, the question is how quickly the marginal cost of installed capacity will rise. The joint study team estimates that on average, the levelized cost of hydroelectric generation will rise to at least 0.30–0.35 yuan/kWh by the year 2020 for large-scale projects under the baseline alternative energy scenario. Since more than 70 percent of China's hydro resources are concentrated in remote regions of southwest China, where both construction and distribution costs are higher, it is likely that an expansion of hydro capacity beyond the baseline will result in levelized costs at or above 0.35 yuan/kWh. While a number of the new hydroelectric schemes may still be economically attractive because of system regulation and peaking capabilities, their costs are still well above the estimated levelized costs of coal for baseload generation.

Nuclear power. Many Chinese energy experts argue that the development of nuclear power is inevitable given the perceived lack of other large-scale alternatives and the difficulties that China will have in transporting coal from inland mines to high-demand coastal areas, especially in the Southeast. There are vast differences of opinion surrounding the costs and expansion capabilities of the nuclear power industry in China. Chinese nuclear proponents anticipate that capital costs will fall to 6,500 yuan/kW (US$1,180/kW in 1990 US$) by the year 2020 as China develops its own nuclear production industry. However, even with such low estimates, the levelized costs of nuclear power are 40 percent above the high estimates for modern coal-fired baseload generation in 2020. If the cost estimates of the international experts participating in this study (US$1,900–2,700/kW) prove more accurate, nuclear power would be too expensive to compete with coal in China.

Wind. Wind turbine power generation is one of the most proven renewable energy technologies worldwide and is developing rapidly. In California, wind generation is among the least-cost options for electricity production. Wind farms in the United States in the 50 to 100 MW scale are being installed for as little as US$700/kW. At this cost, wind generated electricity would be approximately 0.04 $/kWh, which could be competitive with coal under conditions of equal reliability. Chinese experts estimate that wind-powered electricity generation would cost in the range of Y7,000–10,000/kW (US$1,490–2,130/kW). One factor increasing the current cost of wind farms in China is size: installations are smaller and Chinese-manufactured wind turbines are not as big

---

[12] Cost estimates for coal-bed methane recovery and use were obtained from the Global Climate Change Division, U.S. Environmental Protection Agency, which is involved in a GEF-supported technical assistance project assessing coal-bed methane use in China.

[13] This estimate is based on the Yangzhou thermal power project in Jiangsu, financed in part by the World Bank. The coal plant at Yangzhou includes both domestic and foreign-produced equipment and includes state-of-the-art environmental controls, including electrostatic precipitators for particulate control (more than 99 percent removal efficiency) and low-NOx burners. Although there are no special SO_2 controls, the plant will burn very low-sulfur (0.3-0.4 percent) coal.

[14] The high end of the levelized costs is based on the higher capital costs and increased costs for delivered coal, which could result from a rise in marginal transport costs if more coal has to be moved.

as foreign designs (20–350 kW). A second reason for the high costs of wind generation in China is that the cost of building backup facilities has been included. The key to lowering the installed costs of wind turbines in China in the future is to improve domestic turbine designs to make them larger and more reliable, to capture scale economies by establishing larger wind farms, and to select sites where wind regimes and power system characteristics minimize requirements for backup capacity. At the low end of the range for installed costs, wind power is likely to be competitive with coal-fired power generation.

Natural gas. Electricity generation using natural gas is a proven technology that is cleaner than coal-based options, and, when used in a combined-cycle system, can be competitive with coal for baseload generation on a cost per kilowatt-hour basis. In addition, a combined-cycle gas system emits 60–70 percent less CO_2 per unit of electricity than a coal-fired plant. The costs of generating electricity with natural gas in China are highly dependent on the source of the gas. The cheapest sources are the domestic on-land reserves close to consumption centers. However, proven gas resources in China are very limited compared with both coal and petroleum. Moreover, for both domestic and imported gas, it is not clear that electric power generation is the highest-valued use of natural gas in China. If domestic natural gas is to replace coal for electric power generation in China, vast new domestic reserves must be found. Some additional gas may be available through importation, either in the form of LNG or by pipeline from Russia or Central Asia. The high capital costs for developing LNG, competition for gas supplies in the Asia region, and limited number of suitable ports are likely to restrict China's use of LNG in the medium term. Nonetheless, levelized

costs for LNG-generated electric power in other Asian countries are currently as low as 0.05 $/kWh, which would make it one of the lower-cost alternative sources of energy for power generation in China, particularly in areas of the country far from coal reserves, such as the southeast coast.

Solar. Technical improvements in photovoltaic (PV) electric generation over the past twenty years have resulted in a steady decrease in installed costs. As a result, solar photovoltaics have become economic in remote locations in developed and developing countries. Still, the cost of electricity from solar PV is more than four times the cost of conventional power generation cost levels in China and the United States. Additional cost reductions will require that cell production and assembly move from small batch, manual production to fully automated manufacturing. Even with the most optimistic of assumptions for the next century, experts expect that PV-generated electricity in China will be twice the cost of coal-generated electricity. Solar thermal power station proponents expect greater efficiency from mass production of heliostats and reduced receiver costs. To realize these efficiencies will require a more aggressive development strategy than is currently being pursued in China or in other countries.

ALTERNATIVE SOURCES FOR DIRECT USE

In addition to coal, China currently uses oil, natural gas, and biomass fuels as direct energy sources. Under the Baseline GHG Scenario, China will require about 2 billion tce in the year 2020 for non-power energy applications, including transport, residential cooking and heating, and industrial process heat. It is assumed in the analysis that petroleum will be used

Table 3.4 Energy and electricity production by energy type, 1990 and 2020

	1990				2020, baseline				2020, high-substitution			
	Total primary energy (mtce)	%	Electricity generation (TWh)	%	Total primary energy (mtce)	%	Electricity generation (TWh)	%	Total primary energy (mtce)	%	Electricity generation (TWh)	%
Coal	752	76	432	70	2,220	67	2,913	76	1,915	58	2,249	58
Oil and gas	184	19	62	10	783	24	83	2	750	23	106	3
Hydro	51	5	127	20	209	6	601	16	250	8	719	19
Nuclear	0	0	0	0	72	2	208	5	198	6	568	15
Other	0	0	0	0	16	0	45	1	187	6	208	5
Total	987	100	621	100	3,300	100	3,850	100	3,300	100	3,850	100

Sources: China Statistical Yearbook (1990); China GHG Study (2020).

almost exclusively in the transport sector. Fuel for direct use in the residential and industrial sectors will include coal, gas, and renewable energy such as solar and fuelwood.

Natural gas. Because of its convenience, natural gas is highly valued in China as a residential cooking fuel. Given its cleanliness, natural gas will become even more important as a substitute for coal for environmental reasons, particularly in urban areas. If environmental benefits are considered, the highest-valued use of natural gas is expected to be in the residential sector, and, depending on the amount of domestic gas that is discovered, may preclude much natural gas being used by the power sector.

Coal-bed methane. Methane from coal mining is one of China's best alternative energy options and can be developed in the near term. Economic analysis of various uses of coal-bed methane carried out under a GEF technical assistance study[15] concludes that the highest valued use of the gas is as a substitute for coal use by coal mines and other consumers in the immediate locality. Other uses, including piping of the gas to nearby cities for residential consumption and on-site electricity generation, also show rates of return above 12 percent at current market prices for gas, coal, and electric power. The main barriers to the development of coal-bed methane have been low gas prices and a lack of technical expertise in gas re-

covery and processing. The reform of natural gas prices in China since 1990 has led to an increase in the amount of coal-bed methane recovered. This trend is likely to continue given the local environmental benefits of gas use and residential consumers' willingness to pay for gas.

Fuelwood. Unlike most other alternative energy sources, the costs of producing fuelwood in China can be lower than coal on an energy equivalent basis. Based on an economic analysis of fuelwood plantations in which fuelwood is grown under good conditions and intensive management, the discounted cost of producing a ton of fuelwood was found to be 119–245 Y/tce compared with coal at about 160 Y/tce. Fuelwood production costs are lowest in South China, where costs range from 119–141 Y/tce. However, the current price of fuelwood sold by state forest farms in China is only about 90 Y/tce, which is below the break-even price for all fuelwood plantations. If fuelwood were used in China for more commercial purposes, such as tobacco and tea drying or power generation, fuelwood plantations would have a ready source of funds for development and a true market price for fuelwood could develop. However, while strong commercial demand would aid in the development of fuelwood plantations, Chinese foresters argue that strong demand and higher prices make it even more difficult to protect natural forests from illegal felling. Protection of forests has been the major reason for governmental support of fuelwood plantations.

---

[15] *Coal-bed Methane in China*, GEF Technical Assistance Study.

Table 3.5 China: Estimates of capital and levelized costs of electric power, 2020

	Investment cost (1990 Y/kW)	Levelized cost (1990 Y/kWh)
Coal	4,000–5,000	0.18–0.25
Wind	3,300–6,500	0.20–0.37
Geothermal	7,000–14,000	0.28–0.45
Incremental hydro	6,000–7,500	0.30–0.35
Nuclear	7,000–12,000	0.35–0.66
Biomass (gasified)	5,000–10,000	0.29–0.45
LNG	4,000–8,000	0.24–0.40
Solar PV	10,000–20,000	0.50–1.00
Solar thermal	8,000–12,000	0.32–0.70

Source: Alternative Energy Supply Options to Substitute for Carbon-Intensive Fuels, December 1994 subreport prepared for the China GHG Study.

Other renewables. In rural areas not connected to an electric power grid, wind can be a low-cost source of power for agricultural, residential, and light industrial purposes. Solar energy, both active and passive, could be an important source of residential water heating and space heating and cooling. There are reportedly about 2 million residential solar water heaters in China today and another 500,000 domestically produced units are being sold annually. Although precise cost information on solar water heaters produced in China is not available, domestic units are likely to be competitive in certain parts of China, and with further development, their use could be expanded in the residential and commercial sector and in some industrial process heat applications. One of the drawbacks of solar and wind energy which increases their cost is that backup capacity may be needed, particularly for commercial and industrial applications.

GHG CONTROL IN THE FORESTRY SECTOR

Sequestration, or the storage of carbon in woody biomass and soils, can be a cost-effective means of reducing net GHG emissions in developing countries.[16] The key to low-cost carbon sequestration in developing countries is to take advantage of the financial and social benefits of forestry development. Under the right conditions, multi-use afforestation projects, the planting of timber and fuelwood plantations, and the management of open forests in China can yield positive financial returns, meaning that the cost of carbon sequestration from these projects is low.

China's forestry sector

FOREST RESOURCES

China is deficient in forest resources. It has 131 million hectares of forested land, but forests account for only 13.6 percent of the land area. Forest resources amount to only 0.11 hectares per capita, less than one-sixth the world average. The standing volume of wood is about 11 billion cubic meters (m^3), equivalent to about 9.5 m^3 per capita, far below the world average of 66 m^3 per capita. Natural forests account for about three-quarters of the forested area in China and 95 percent of the standing wood volume. Most natural forests are located in the northeastern prov-

inces of Jilin, Liaoning, and Inner Mongolia, and in Sichuan and Yunnan in the southwest. By the late 1980s, the stock of mature and over-mature forests amounted to about 2.6 billion m^3, of which only about half was commercial. At current rates of exploitation, the stock of commercial forests will be exhausted within ten years. China is the third largest consumer of forest products in the world, and the demand for wood products has been expanding with the development of the economy. Current wood consumption is approximately 300 million m^3 per year, of which about half is industrial roundwood, a third is fuelwood, and the remainder is used for rural construction. During the 1980s, the consumption of forest resources exceeded the annual increase in growing stock by some 20 million m^3.

Because of the shortage of forest resources in China and the gap between the supply and demand for wood products, China has instituted the largest afforestation program in the world. Between 1984 and 1988, China planted an average of 3.25 million hectares per year; the net increase in forested land was about 0.65 million hectares per year. In 1989 and 1990, the area of planting exceeded 5 million hectares per year, with the net increase surpassing 2 million hectares per year. In addition to timber and fuelwood plantation forests, China has initiated a large-scale protective afforestation program in North China and has designated some 3 million hectares of open forests, mostly on steep slopes, to be protected for forest development. If China continues to increase its forested area at the rate of the past five years, forests will cover more than 15 percent of its land area by the end of the century.

FGHY PROGRAM

In the past, Chinese plantation forests have had low survival rates and low productivity, largely because of poor management and inferior lands. Since the mid-1980s a new, fast-growing, high-yield (FGHY) afforestation program has been initiated in China. This program relies on (i) planting on good quality sites, (ii) improved genetic materials and seedling preparation, and (iii) detailed silviculture prescriptions for planting, tending, and harvesting. Tree species are selected that can meet minimum growth rates under given soil and climatic conditions. Among the most common FGHY species in China are larch, fir, poplar, pine, eucalyptus, and paulownia. More than a quarter of the plantations established in China since 1985 have been FGHY plantations.

---

[16] Net GHG emissions is the sum of all emission sources less the amount of carbon that can be captured and stored in plant biomass or soils.

BARRIERS TO LARGE-SCALE FGHY
FORESTRY DEVELOPMENT

While there are approximately 200 million hectares of land theoretically available in China for forestry development, there are competing claims for these lands, including agriculture, animal husbandry, and urban and industrial development. Moreover, much of the land available for forest development in China is of poor quality and inappropriate for intensive forestry plantations. Some types of forestry development, including fuelwood plantations and open forest management, tend to be financially viable only in southern China, whereas much of the land available for forestry development is in the north and northwest.

Forests in China are managed by state forest farms, collectives, and individuals. Although most land is controlled by state forest farms, much of this land is contracted to collectives and individual households, which provide labor and inputs. The quality of management by collectives and individuals is generally low, since they have had limited access to advanced silviculture techniques and since most face financial constraints. In the 1980s, roughly a quarter of the total forested land was transferred from the state to households as part of the household responsibility system. While land tenure relations were improved, household forest farms continue to be characterized by low productivity and minimal adoption of improved silviculture techniques.

Modeling carbon sequestration in China

SCENARIO ASSUMPTIONS

Two computerized models were developed for analyzing the magnitude and costs of carbon sequestration in China: (i) a national model that predicts standing volume, age class structure, and forest carbon balance; and (ii) plantation models for assessing the costs and benefits of forestry planting and net costs of carbon sequestration. Both models are simulated for thirty years, beginning in 1990, with the planted or managed areas based on the scenarios given in Table 3.6. The medium scenario roughly corresponds to the actual level of planting in China during the early 1990s.

NATIONAL FORESTRY MODEL

The national forestry model was used in Chapter 2 to estimate the potential for carbon sequestration

in China's forests. The model divides forested land into five age classes (Table 3.7). Forest growth is represented by the movement of land through these age classes, while standing volume is calculated by multiplying the area in each age class by the average stand age and the mean annual increment (m^3/ha/yr). An algorithm based on demand projections determines the area of timberland harvested each year, with harvesting beginning in the oldest age class and working its way downward; the youngest age class is prohibited from being harvested. Soil carbon absorption follows the S-shaped pattern of tree growth, with the volume of soil carbon equal to the difference between the initial and equilibrium carbon levels.

The planting and reforestation program evaluated by the model shows that China's stemwood will more than double between 1990 and 2020, from 12 to 28 billion m^3. However, the harvesting of old growth forests and the planting of new forests will significantly change the age structure of China's forests. Currently, mature and over-mature stands make up about 20 percent of China's timber reserves. After thirty years, only 6 percent of its reserves are projected to be mature and over-mature stands, demonstrating a dramatic decrease in old-growth forests and foreshadowing shortages of large-diameter timber and a loss of old-growth habitat. A shortage of large-diameter timber in coming decades means that China must either increase the volume of timber imports or improve its capability to manufacture specialty wood products (e.g., medium density fiberboard and laminated beams) from small-diameter timber.

FORESTRY PLANTATION MODELS: FINANCIAL AND ECONOMIC ANALYSIS

Several studies of the costs of carbon sequestration through tree planting and modified forestry practices have reinforced the perception that the net costs of carbon sequestration by the forestry sector are positive.[17] Such studies have generally calculated the cost of carbon sequestration as the total cost of tree planting—purchasing or renting land, establishing and maintaining trees—divided by total carbon sequestration (in biomass and soil). Not generally considered are the direct economic benefits from tree planting; that is, revenues from the sale of timber products (sawlogs, pulpwood, and fuelwood), social ben-

---

[17] See for example, R.J. Moulton, and K.R. Richards, *Costs of Sequestering Carbon Through Tree Planting and Forest Management in the United States* (Washington, DC:U.S. Department of Agriculture Forest Service, December 1990).

Table 3.6 Planting assumptions (million hectares per year)

Scenario	Replanting of clear-cut forests	Incremental plantations	Incremental natural regeneration (open forests)	Total area
Low	2.62	1.34	0.89	4.85
Medium	2.62	2.01	1.34	5.97
High	2.62	2.68	1.79	7.09

efits from reducing deforestation, and environmental benefits such as erosion control and habitat protection. While the regular thinning and harvesting of timber plantations can result in lower overall carbon sequestration than if the forests are left untouched, the financial benefits from the use of forestry products are crucial to the success of afforestation programs, particularly in developing countries.

This analysis considers the net cost of carbon sequestration from tree planting and modified forestry practices in China, whereby the private financial benefits are subtracted from the costs. Financial tables, which include the costs of establishment, maintenance, harvesting, and reforestation, and the benefits from sales of sawlogs, pulpwood, and fuelwood, have been prepared for intensive, extensive, and fuelwood plantations, and for the improved management of open forests.[18] To calculate the cost-effectiveness of carbon sequestration, the net present value of each respective project is divided by the discounted carbon sequestered in living biomass and in the soil.

For the financial analysis of various plantation programs, the scenarios of new "incremental plantations" (Table 3.8) were subdivided into three categories: (i) intensively managed plantations, which refers to the planting of fast-growing species, such as Chinese fir (*Abies*), masson pine, larch (*Larix*), paulownia, poplar, and eucalyptus on good quality land to maximize biomass production and financial returns, (ii) extensively managed plantations, which refers to the planting and management of naturally occurring species in China, such as oak, birch, spruce, fir, and various pine species, which would be managed as commercial forests but with much longer rotations and lower costs than intensive plantations; and (iii) fuelwood plantations, which refers to the

planting of fast-growing regenerable species, such as locust and eucalyptus, with the intention of maximizing biomass production and financial returns. The area of fuelwood plantations (medium scenario) is taken from the 1991–1995 five-year plan target (2.6 million hectares, or 520,000 ha/yr), while the proportion of intensive and extensive plantations is based on the current situation (40/60) as reflected in government forestry statistics. In 1990, 3.67 million hectares were closed for natural regeneration; this is referred to here as open forest management.

RESULTS OF THE FINANCIAL ANALYSIS

Intensively managed plantations. Intensively managed plantations would be established on the most productive forestry lands and have the highest growth rates and the highest costs. Based on the most recent (1993) prices for timber products, and even assuming the substantially lower prices in 1990, ten of the twelve intensive plantation species yielded internal rates of return (IRR) above the target rate of 12 percent. The World Bank recently appraised similar FGHY plantations as part of a forestry project in sixteen provinces in China and estimated the average rate of return for such plantations at 18.1 percent.[19]

[19] The World Bank, Staff Appraisal Report, *China: Forest Resource Development and Protection Project*, May 1994.

Table 3.7 Standing stock by age class (million m³)

Age class	1990	2000	2010	2020
Total	11,960	16,140	20,490	27,740
Young	970	2,834	4,630	5,948
Middle-age	4,345	5,152	6,359	11,130
Premature	2,127	4,573	6,773	8,988
Mature	3,503	3,425	2,623	1,607
Over-mature	1,015	156	105	67

Source: National forestry model, joint study team.

[18] Financial data for the intensive plantations is taken from the National Afforestation Project, which has received financial and technical assistance from the World Bank.

Extensively managed plantations. While the costs of extensive plantations are much lower than intensive plantations, so are the growth rates. The financial analysis shows that only three of the ten extensive plantation models yielded rates of return above 12 percent. The models that yielded positive net present values were Chinese fir, masson pine, and Yunnan pine growing in South and Southwest China.

Fuelwood plantations. The Chinese government has promoted fuelwood plantations to reduce the adverse impact of fuelwood collection on natural forests and timber plantations. Nearly two-thirds of fuelwood production would come from fuelwood plantations, and one-third from thinnings and removal of dead or dying wood from other plantations. In contrast to other plantations, the production of fuelwood is not financially viable at official prices on current state-owned fuelwood plantations (30–50 Y/t). However, the true market price or opportunity cost of fuelwood is probably double this amount based on the cost of collection or compared with other fuels. At 100 Y/t, which is roughly the fuel equivalent price of coal in the regions reviewed minus the difference in fuel combustion efficiency, fuelwood production exceeds the 12 percent rate of return target in South and Southwest China.

Open forest management. Growth rates of existing open forests can be improved by applying better silviculture techniques and by protecting the forests against destructive felling and encroachment. Some timber and fuelwood would be available from these lands through selective thinning and removal of dead or dying trees. The financial analysis suggests that improved open forest management is commercially viable only in South China.

Sensitivity analysis. The variable that most influences the rate of return of the plantation projects is the price of forest products (sawlogs, smallwood, pulpwood, and fuelwood). The most recent price information for fast-growing species in intensive plantations, collected as part of the National Afforestation Project, has been used in the plantation models. Over the past five years, timber prices throughout the country have increased substantially as the government has eased price controls on "in-plan" timber harvests and as a growing percentage of "above-plan" timber harvests have been sold at market prices.

Conclusions

The net cost analysis demonstrates that the least-cost means of sequestering carbon in China is through FGHY timber plantations. While not nearly as attractive as FGHY plantations, other low-cost options for sequestering carbon in China include planting fast-growing natural species, such as Chinese fir and pine, on extensive plantations in South China, and open forest management, also in South China where growth rates are high. By contrast, gross cost analyses done in the United States and elsewhere conclude that least-cost sequestration forestry projects are those that minimize total costs, even when the forestry projects in question are not commercial.[20]

---

[20] For instance, Xu calculated the gross costs of carbon sequestration for China using the Moulton and Richards methodology and concluded that the least-cost measure was management of open forests in Southwest China, where the mean annual increment was high and the costs of labor and other inputs were low. By contrast, under the net cost approach, open forest management does not meet the target rate of return (12%) in most regions of the country and in no region is open forest management as attractive in financial terms as intensive plantations. Xu Deying, *Economic Analysis and Forestry Options for Mitigating Global Climate Change: A Chinese Case Study*, Research Institute of Forestry, Chinese Academy of Forestry (Beijing, China, 1993).

Table 3.8 Plantation scenarios (thousand hectares per year)

Scenario	Intensively managed plantations	Extensively managed plantations	Fuelwood plantations	Total incremental plantations
Low	280	710	350	1,340
Medium	420	1,070	520	2,010
High	538	1,360	780	2,680

Source: Joint study team, China GHG Study. See Table 3.6 for total planting.

GHG CONTROL IN THE AGRICULTURAL SECTOR

There are several agricultural technologies and practices that can reduce GHG emissions and that are being promoted in China for non-GHG reduction reasons. Research on new and exotic agricultural techniques for reducing GHG emissions is being conducted in China and abroad and may yield additional low-cost options for emissions reduction. However, this section focuses on currently available no-regrets options because they are the most likely to be implemented in the short term and will not require large government subsidies. The analysis covers methane emissions from rice cultivation and from large ruminant animals, which together accounted for about 90 percent of China's agricultural sector emissions in 1990.[21]

Reducing methane emissions from ruminant animals

In 1990, ruminant animals accounted for more than a quarter of GHG emissions from the agricultural sector in China. Improved feed and breeding programs for large ruminant animals have the potential to reduce methane emissions by increasing the efficiency with which products (meat, milk, and draft power) are produced. Based on rough estimates made in 1990, the average animal in China yielded only about 5.5 kg of meat per head and about 1,500 kg of milk per lactation, compared with 130 kg of meat per head and 7,000 kg of milk per lactation in the United States.[22] Economic analysis of two popular animal husbandry programs in China shows that methane emissions can be reduced by 25–50 percent with no net costs; that is, both projects have sound financial and economic rates of return.

Improved breeding. The efficiency of beef and milk production can be improved in China by selectively breeding exotic species of cattle and buffalo with domestic species to improve genetic quality. In 1978, the Chinese State Science and Technology Commission decided that artificial insemination should be used to improve domestic breeds of cattle. In 1980,

[21] For details, see the subreport prepared for this study, *Greenhouse Gas Control in the Agricultural Sector*, September 1994.

[22] Sollod and Walters, *Reducing Ruminant Methane Emissions in China*, U.S. Environmental Protection Agency, Global Change Division, Office of Atmospheric Programs (Washington, DC: October 1992), p. 3.

140 bulls were imported from North America and Western Europe for use in the program. Since 1980, Indian and Pakistani milk breeds such as Mara and Ravi Nili have been bred with Chinese buffalo in South China to increase milk production, while Simmental and Holstein-Frisian species have been bred with Chinese yellow cattle to improve both milk and meat production.

Artificial insemination using frozen semen has several advantages over natural service. For example, it can prevent premature mating, fix the dates of breeding, and increase the conception rate. Another advantage of frozen semen artificial insemination is its ease of mastery. With about fifteen days of training, a veterinarian can learn the skill and produce a conception rate of 50 to 80 percent.

From a commercial beef producer's perspective, the artificial insemination program has been a financial success. Case study results indicate a 74 percent financial rate of return from using improved cattle instead of domestic yellow cattle. Sensitivity analysis given alternative weight gain and feed consumption levels indicates stable results.

Ammoniated feed. The treatment of straw and other crop residues with various forms of ammonia (such as urea) allows otherwise low-nutrient residues to become an important component of an animal's diet. This reduces feed costs and thus increases the profitability of raising animals for meat, milk, or draught power. In addition to raising rural incomes and helping to expand meat and milk production, this technique reduces rural air pollution by limiting crop residue burning. Large-scale trials using ammoniated feed and protein supplements for cattle have been carried out in Henan and Hebei provinces under a cooperative project between provincial institutions, the Ministry of Agriculture, and UNDP. The trial stage of the project has been successfully completed and the Chinese government plans to expand the ammoniated feed program to other parts of the country by guaranteeing the supply of urea and promoting rural credit for cattle and equipment purchases.

In the early 1980s, China started a pilot program of using ammoniated straw to feed cattle. Straw and stalks are ammoniated by cutting them to 2–3 cm lengths, mixing them with urea and water, and allowing them to ferment. According to estimates, the ammoniated straw contains two to three times as much crude protein as the raw product, and the amount of consumption and speed of digestibility increase by 20 percent.

After a slow start, the program has recently expanded rapidly. In 1986, only 43,000 tons of straw were ammoniated while in 1991 the amount had grown to 3.71 million tons. Reasons for this rapid adoption include low investment costs, ease of understanding and mastering the technique, availability of raw materials, and large financial returns.

A primary factor in the attractive financial returns is enhanced weight gain. Research at the Hebei Animal Husbandry Research Institute found average daily weight gains of 644 grams when treated straw was used versus 348 grams with untreated straw, an 85 percent improvement. In both cases, treatments included 1.5 kg of cotton seed cake and a mineral supplement. Using these weight gain differentials, a financial rate of return of 110 percent is obtained when a hypothetical project using ammoniated feed is compared with a similar project using untreated straw.

Rice production

Although economic analyses of rice cultivation practices for reducing methane emissions were not conducted for this project, the Chinese Ministry of Agriculture reports that the programs discussed below are being adopted in various parts of the country for other reasons. To the extent that farmers are adopting these techniques without government assistance, it can be assumed that the financial benefits of these techniques outweigh the costs and that the corresponding costs in terms of methane reduction are zero.

Anaerobic decomposition of organic material in flooded rice fields produces methane, which is released to the atmosphere primarily by transport through the rice plants. A major determinant of the amount of methane produced is the level and duration of flooding. For this reason, cultivation techniques that limit the flooding of fields may reduce methane emissions from rice production.

Semi-dry rice cultivation. Semi-dry cultivation of rice involves digging furrows and flooding them two-thirds full. Rice is transplanted on the ridge, promoting development of the root system. The lower water level reduces the extent of anaerobic decomposition of organic material and thus the production of methane. A pilot project for semi-dry cultivation of rice was begun in 1982 in Sichuan Province and was introduced in several other southern provinces

between 1985 and 1988. By 1988, 674,000 hectares—2 percent of China's total rice area—was cultivated in this manner. The technique has been combined with fish farming and the inter-cropping of aquatic vegetables and medicinal plants to further raise rural incomes. Although the technique saves water, it is very labor intensive, which may discourage its implementation.

Intermittent irrigation of paddy fields. Rice shoots do not have to live in water all the time. At certain times the soil needs to be moist, while at other times the field can be allowed to dry, which promotes soil ventilation. The appropriate timing of moist versus dry periods can improve productivity and reduce methane emissions. Prolonged flooding enables peak anaerobic activity, which produces methane. By shortening the period of flooding, the conditions for growth of methanogenic bacteria are disrupted, and methane production is suppressed.

In recent years, this production technology has been widely adopted in the high-yield rice growing areas of China. However, there are limits to its use. Regions with water shortages and low-lying land that is slow to drain cannot benefit from this technology. As a result, it is estimated that by 2020, intermittent irrigation in China will account for no more than 10–15 percent of production.

Chapter 4

Conclusions and Recommendations

GUIDING PRINCIPLES

China currently accounts for about 10 percent of global CO_2 emissions and this percentage is almost certain to increase as China's economy develops. Given the prospects for continued rapid economic growth in China well into the next century, it is essential that China be included in any international strategy to mitigate global climate change. Many developed countries have already provided bilateral assistance to China to address climate change issues and have endorsed the provision of similar assistance to China through global environmental initiatives such as the Global Environment Facility (GEF). This study, the first country-specific study to be funded by the GEF, and the expected follow-up investment and technical assistance projects, demonstrate the importance that the Chinese government and the international community attach to reducing GHG emissions in China.

It is also in China's own interest to limit GHG emissions. While small countries can do little to affect global climate change since their emissions are a minuscule part of global GHG emissions, China can have a substantial effect on incremental global emissions and thus can limit the potential climate change impacts that will affect China and the rest of the world. Furthermore, climate change impacts are likely to be more severe for China and other low-income countries since a larger share of their income is derived from climate-sensitive sectors such as agriculture and because they are least able to afford adaptation or avoidance measures. Therefore, China has a strong incentive to take action to reduce GHG emissions, regardless of the potential impact on other countries.

Despite the imperative to limit GHG emissions, economic development and poverty alleviation must be the top priority for China and other low-income countries. China's development strategy will have an enormous impact on the level of GHG emissions. The most important and cost-effective way of reducing GHG emissions without impairing development is to improve the efficiency of the economy. This can best be accomplished by continuing and deepening the economic reform program that China initiated in 1978.

In addition to improving the overall efficiency of the economy, there are several low-cost options for reducing GHG emissions in China over the short to medium term (before 2010). As shown in previous chapters, the GHG reduction potential of these projects is significant. Moreover, because they provide other substantial benefits in terms of financial, economic, and, in many cases, local environmental returns, there is little or no net cost for GHG reduction. Such "no-regrets" projects should be the primary focus of China's GHG reduction strategy over the next 10–15 years.[1]

While economic reform and no-regrets projects should form the basis of China's GHG reduction strategy for the immediate future, the only long-term GHG reduction option for China and the world is the development of alternatives to fossil fuel. Some alternative energy technologies are commercial in China today and should be expanded as part of China's medium-term no-regrets strategy. However, many of the most promising non-carbon energy sources cannot be adopted on a large scale in China until technologies are further developed that can compete in both scale and cost with coal. For this to happen, China must enlarge its development program for alternative energy technology, substantially increasing current development investments to make a large impact as early as possible.

---

[1] For a discussion of the concepts of no-regrets and net costs of GHG reduction, see p. 34.

In sum, China's strategy for reducing GHG emissions should be based on the following principles:

• Continue and expand the economic reform program to improve the efficiency of resource use;

• Accelerate the implementation of "no-regrets" projects over the short to medium term; and,

• Enlarge and improve the program to develop low-carbon-intensive energy technologies for the longer term.

RECOMMENDED STRATEGY FOR REDUCING GREENHOUSE GAS EMISSIONS IN CHINA

Based on the detailed analysis conducted for this study, the joint study team recommends that China pursue a two-pronged strategy for reducing greenhouse gas emissions: (i) further promote economic reform, whereby market incentives and regulatory controls are adjusted or introduced to improve resource allocation and encourage energy efficiency; and (ii) implement a set of priority investment and technical assistance programs to accelerate the adoption of more efficient and low-carbon technologies and improve the institutional and human resource capacity necessary to implement and sustain these programs. This strategy for global climate change mitigation is largely consistent with China's domestic environmental and economic modernization objectives, and thus does not pose a major new or different set of conditions or constraints on the public sector or enterprises.

Economic reform

The proportion of China's GHG emissions from energy consumption—currently about four-fifths—is expected to increase in the future.[2] China's GHG reduction strategy, therefore, must aim to maximize the economic value derived from energy consumption through improvements in the efficiency of energy use per unit of economic output. As discussed in Chapter 2, improvements in energy efficiency are achieved both indirectly, through changes in the structure of the economy, and directly, through the use of improved technologies that result in energy savings per unit of physical output. The following structural changes will result in declines in the energy-intensiveness of the economy:

---

[2] See p. 31.

• changes in the share of energy use and output among various sectors (e.g., agriculture, industry, and services);

• changes in the share of different industrial subsectors in industrial output; and

• changes in the product mix and sources of value added within various sectors.

In China, there is vast potential for energy-efficiency gains from increases in the value of the industrial product mix per unit of energy input through (i) increasing product diversification and specialization and (ii) improving product quality. The potential for gains in energy efficiency is greater through indirect structural changes than through technical efficiency improvements (see pp. 23-24).

The future magnitude of indirect energy-efficiency gains through structural change is tied to the speed and depth of progress in China's economic system reform. This includes the related issues of increasing the importance of market forces, increasing competition, and increasing enterprise autonomy and accountability. Progress on economic system reform will also help catalyze technical energy-efficiency initiatives, as discussed on pp. 59-60.

ENTERPRISE REFORM

Reforms must be completed to make enterprises economically autonomous units that are fully accountable for their own profits and losses. Completion of such reforms will further encourage the cost-consciousness, profit-seeking innovation, and product mix changes necessary to support new gains in both indirect and direct energy efficiency. Until recently, enterprise decision making in China was closely tied to government policy, which often determined product mix, scale of production, capital investment, and factor payments. Since the mid-1980s, state-owned enterprises (SOEs) have been given greater control over their operations and there has been a rapid expansion of "nonstate" entities, including township and village enterprises (TVEs), and private firms. Nonstate entities have generated new employment, contributed to economic production and exports, and there is growing evidence that they have had a positive impact on the competitive market environment for all enterprises in China. Further restructuring of SOEs is needed to improve economic efficiency. This includes: (i) reducing and eventually eliminating the direct and indirect subsidies that many SOEs still enjoy, and enforcing true accountability through a "hard budget constraint;" and (ii) transferring, to independent enterprises or the government, the SOEs' traditional responsibility for providing a broad range of social services—housing, retirement pensions, education, and health care.

INCENTIVES AND REGULATORY FRAMEWORK

Historically, below-market energy prices were a primary reason for high per-unit energy consumption and are one of the reasons Chinese enterprises have not been motivated to invest in energy conservation. Over the last decade, China has made great progress in reforming energy prices. Oil product prices are now close to world market levels, average end-use electricity prices now compare reasonably well with long-run marginal costs, and, as of 1994, coal is sold at market prices. Nevertheless, further reform is still necessary. Measures to further rationalize energy prices include adoption of nondiscriminatory pricing of natural gas; improvements in the structure of electricity prices, including elimination of subsidies and low in-plan prices for certain industrial consumers; and adoption of differential rates for coal to better reflect quality. These measures would help to improve overall energy efficiency by Chinese consumers.

Other policies that distort enterprise behavior and result in inefficient resource use and the consequent excess of GHG emissions include (i) regulatory and trade barriers that restrict market entry and exit for new competitors (both domestic and foreign), (ii) tax policy that discriminates against enterprise ownership forms, and (iii) foreign trade policies that restrict the import or transfer of certain goods.

FACTOR MARKET DEVELOPMENT

The development and continued reform of China's financial system are necessary to ensure that all enterprises can borrow against future earnings to respond to market conditions. The lack of a functioning capital market and government restrictions on the type of investments allowed were why Chinese enterprises historically did not operate at proper scale, invest in new plants and equipment, or move into more socially desirable product lines. Several improvements have been made in China's capital markets since the early 1980s, including the introduction of bank loans to replace state budgetary appropriations, separation of the governmental and commercial functions of the People's Bank of China, and

expansion of the stock and bond markets for financing enterprise investment. Further reforms are needed in China's capital market and financial sector to ensure that all firms have equal access. In addition, mechanisms for writing off bad loans from ailing SOEs and improved allowances for bankruptcies are necessary to ensure that capital market reforms continue.

A functioning labor market is essential to the efficient operation of enterprises in China. Because SOEs have historically provided lifetime employment and numerous social services, they have not been able to respond to changing market conditions. The output from TVEs and private firms has expanded rapidly in China over the past decade in part because of their enhanced ability to acquire or shed labor.

Access to and the development of advanced technologies is important for the modernization of China's industrial sector and, consequently, for the reduction of per-unit energy consumption. The reduction of trade barriers, the establishment of patent protection, and the mobility of scientific personnel are all necessary to ensure that internationally advanced technologies are available. In addition, the Chinese government may need to ensure that technologies important for the global environment are developed. For instance, government support of research and development for alternative energy technologies is important for both strategic and environmental reasons, and, without such support, enterprises will underinvest in such research and development.

ENVIRONMENTAL REGULATORY POLICY

Environmental regulation can be important for reducing GHG emissions in China by (i) encouraging the adoption of cleaner and more efficient new processes and technologies, and (ii) making global climate change mitigation an explicit environmental goal. In developed countries, the establishment and enforcement of environmental regulations have been important means of accelerating the acquisition of cleaner and more energy-efficient technologies and equipment. While China has made considerable progress over the past decade in establishing a comprehensive environmental regulatory system, the system must be better suited to a market economy. The expansion of the non-state sector and the granting of greater autonomy to SOEs have reduced the effectiveness of government measures for regulating enterprise behavior, including pollution control. Environmental policies in China should take advantage of the incentives that firms now have to reduce costs and increase profits. Market-based incentives must be provided to firms to encourage them to adopt cleaner and more energy-efficient processes and technologies, while pollution fines should be raised to levels that exceed the cost of prevention or cleanup. In all instances, environmental regulations in China should be more strictly and uniformly enforced. Another aspect of environmental policy, which can be quite cost-effective, is increasing public awareness and participation in environmental pollution control. Publicizing local environmental quality indicators, educating the public about the relative health risks of environmental pollution, and identifying the main polluters and actions needed for compliance are effective measures for reducing environmental pollution.

Internationally, China should enunciate and publicize its strategy for GHG reduction as part of its global climate change initiative. Components of this strategy are contained in this report and in reports already prepared under other environmental plans, including Agenda 21 and the National Environmental Action Plan.

Priority project areas

While overall improvements in economic efficiency through further economic reform are critical for effective GHG abatement, China's GHG control strategy must also include a series of more specific actions. These actions include investment projects, reform of certain sector-specific policies, and efforts to further build institutional, technical, and managerial capacities. In most cases, these actions provide additional benefits and serve other goals and objectives as well.

The joint study team concludes that the project areas described below warrant highest priority for action to reduce GHG emissions in China. This conclusion is based on the analysis in Chapter 2 concerning which measures can contribute most to reducing GHG emissions, and the analysis in Chapter 3 regarding the cost-effectiveness of different measures. Together, these analyses provide an overview of which measures can reduce the most GHG emissions at the lowest cost.

IMPROVEMENTS IN ENERGY EFFICIENCY

Improvement in the direct, technical efficiency of

energy use is clearly the top priority for specific action to mitigate GHG emissions in China over the short and medium term. As shown in Chapter 2 (pp. 24-28), further improvements in the technical efficiency of energy use can have a major impact on China's GHG emissions over the next twenty-five years. As shown in Chapter 3 (pp. 37-38), improvements in energy efficiency also are among the most cost-effective means to reduce GHG emissions, since there are many investments where energy cost savings more than offset total costs. Expanded investment is required to reduce energy use per physical unit of output, both through renovation of existing facilities and through adoption of more efficient processes and equipment in new facilities. Actions to conserve coal are most important because of coal's dominance in the energy mix and the high carbon dioxide emissions associated with its combustion. Industry will continue to be the sector with the greatest potential for energy-efficiency savings. Improving the efficiency of use of certain energy-intensive raw materials, such as steel, requires addressing many of the same problems in implementation as direct technical energy efficiency. These problems can and should be addressed as a component of GHG mitigation.

The challenge for China is to increase and accelerate the levels and effectiveness of investment in energy-efficiency improvements. Continuing enterprise reform and energy price reform have given enterprises more incentives to conserve energy. China also has built a sound institutional network for promoting energy conservation and has developed a wide range of programs, some with marked success. Yet, many cost-effective investments in energy-efficiency renovations have not been implemented. In fact, energy-inefficient designs, processes, and equipment are still widely used in the development or construction of China's new industrial production capacity and housing stock.

Improvements in energy conservation will require the further development of market-based incentives. Timely and effective completion of the transition to a market economy is the most important action that can be taken to encourage enterprises to conserve energy. Successful completion of the reform program to enforce hard budget constraints on SOEs, in particular, is critical to improve energy cost-consciousness in these enterprises. Further efforts should also be made to complete energy price reform (p. 58). Following a review of the impact of the recent liberalization of coal prices, the government should also consider the potential advantages of additional taxation of coal as a means to incorporate some environmental externalities into the cost of coal burning. A first step would be to undertake an in-depth study of the advantages and disadvantages of coal taxation for environmental goals.

To speed progress in meeting global and national environmental goals, the government should supplement its efforts to encourage market forces with more effective initiatives to overcome the barriers impeding progress discussed in Chapter 3. These barriers include insufficient access to information on technical opportunities and experiences, lack of access to foreign technology, technical and market risks, high transaction costs for small investments, and institutional constraints. Government support for such initiatives is especially important now, since enterprises are increasingly interested in market incentives for energy conservation and managers want to expand their experience and knowledge of options. Recommendations for action, especially to support energy conservation in industry, include the following:

Credit facilities. China's credit facilities for energy conservation investment can be useful; however, a series of improvements should be made. Enterprises receiving energy conservation investment financing should be required to demonstrate their overall financial and economic viability or present a credible restructuring plan to achieve such viability. This would ensure that the project is sustainable and that the correct problem is addressed (see p. 41). Given the wealth of investment opportunities with sound life-cycle returns, credit facilities should focus on financially viable investments with longer payback periods, which tend to be less attractive to enterprises and commercial banks.

Concessional finance. Many developed and developing countries have instituted programs to provide tax or duty relief or both, interest rate subsidies, or small grants for selected energy conservation investments. The most effective programs tend to be well targeted, with specific objectives to catalyze larger responses. In China, the best results may be obtained through selective assistance for (i) development and demonstration of new energy-saving technologies, including technology transferred from abroad and innovations carrying substantial technical or market risk; (ii) development and introduction to manufacturing of new, high-efficiency equipment; and (iii) pre-investment and ex-post evaluation work. Adaptation of a program that has been successful in the U.K.,

whereby small grants are provided for innovative conservation investments in exchange for publication of a full assessment of all project results, good or bad, might be attractive in China.

Energy conservation equity investment concepts. The Chinese government should encourage experimentation with new organizational forms being developed abroad. For example, public or private "energy service companies" or "conservation investment companies" finance energy conservation investments in enterprises in exchange for a portion of the life-cycle profits. The companies shoulder the risk by providing a guaranteed operating cost reduction to the participating enterprises.

Improved dissemination of information. Several programs exist at the central and provincial levels for dissemination of information on energy conservation technology, but these do not satisfy the country's enormous needs. There is a pressing need for information on project implementation experiences in enterprises, including full technical details and, most importantly, an assessment of the financial results and cost savings. Programs targeting small enterprises, especially TVEs, need to be developed. Marketing assistance initiatives could aid in the dissemination of high-efficiency technologies, and a government-sponsored energy-efficiency labeling program should be instituted for key types of equipment to report the results of standardized energy-efficiency tests based on common use patterns. Consumer response to all such programs should be surveyed and evaluated, and program revisions incorporated accordingly.

Technical assistance and training. The current need for technical assistance and training exceeds the capacity of the existing institutional network. The government should complete a critical assessment of the network for providing enterprises with energy auditing, preinvestment, and energy-efficiency training services, focusing on the results achieved. It should also identify areas for improvement and follow-up work.

In addition to these efforts to accelerate energy conservation investment, further recommendations on important areas are provided below.

Industrial energy conservation. China's energy conservation agencies are likely to achieve the best results if the efforts to promote conservation investments in industry are targeted toward "classic" industrial energy conservation projects in enterprises that can demonstrate a sound financial future. Examples of such classic conservation projects include waste heat or gas recovery, cogeneration, furnace or kiln renovation, adoption of energy management systems, improved insulation and thermal/steam system renovation, and installation of high-efficiency equipment. Support is needed for projects to conserve energy-intensive raw materials, such as steel. Although large energy-efficiency gains also can be achieved through industrial restructuring projects, assistance for these broader initiatives is best left to other institutions. However, energy conservation agencies can and should help in identifying and assessing the implications for energy consumption of industrial restructuring and modernization initiatives.

High-efficiency energy-consuming equipment. Implementation of a national initiative to develop higher-efficiency small- and medium-scale coal-fired industrial boilers, through adaptation of foreign technology to Chinese conditions, is among the top priorities for reducing GHG emissions over the near term. These industrial boilers account for about 30 percent of current GHG emissions from energy use; cost-effective design improvements could reduce unit coal consumption levels by 10–20 percent. With little potential for exports and profits, there has been a marked lack of international commercial interest in the small coal-fired boiler sector, and Chinese enterprises have little access to foreign technical advances. Without assistance, individual Chinese boiler manufacturers are unlikely to be able to shoulder the costs and technical and market risks associated with upfront outlays for technology acquisition, adaptation, and demonstration, and for trial production of new designs.

Additional support is also needed to encourage the adoption of other types of high-efficiency equipment, such as high-efficiency electric motors, variable-speed motors, and associated high-efficiency industrial electrical equipment; more efficient air-conditioning equipment and refrigerators; efficient lighting devices; and steam traps associated with industrial piping networks. Improving the efficiency of electric motors and associated industrial equipment is most important, since such equipment accounts for almost half of China's total electricity use. Additional support might best include selective assistance for technology development and demonstration, assistance in marketing, provision of better information to consumers including energy-efficiency product labeling where appropriate, and the intro-

duction of mandatory efficiency standards in select cases, such as electric motors and air conditioners. To be effective, efficiency standards must be set so that rigorous testing can be conducted. The standards should be gradually phased in to enable industries to comply without undue hardship and should be strictly enforced.

Improvements in coal processing. Adaptation and dissemination of clean coal technologies will be an important element in China's GHG control strategy. Over the long term, adoption of more efficient coal gasification technology will result in substantial energy-efficiency gains as well as local environmental benefits. For local environmental reasons, China has emphasized expanding the use of coal gas among residential, commercial, and small-scale urban industrial consumers. Development of integrated coal gasification for combined cycle power generation may have advantages in terms of project implementation.

In the near term, however, the most pressing issue relating to cleaner and more efficient coal use is improvement in the quality of the coal supply, through increased washing of steam coal, better sorting and matching of coal varieties and sizes to consumer needs, and briquetting and pelletization. Given that coal is China's dominant source of energy and that more than half of the country's coal is consumed by relatively small-scale users, improvements in the quality and uniformity of supplies can yield large gains in efficiency and reduce carbon dioxide emissions. Yet, in this respect, China has made little progress in the last decade. Improved results depend upon policy support and institution building to (i) complete the reform of the coal supply and distribution system from a supply-driven system to a consumer demand-driven, market-based system, and (ii) improve regulation of air pollution emissions, especially enforcement of air pollution standards (see p. 45).

A recommended means to speed progress is to implement a pilot project to reform the system for coal supply, processing, and utilization in one or two urban areas. Such an initiative could include (i) implementation of a system for rigorous air pollution control regulation and enforcement, (ii) a program to replace the urban fuel supply system with market-oriented coal marketing companies, (iii) targeted investment in coal selection and briquetting facilities, and (iv) selected assistance to consumers to adopt more efficient coal use technology.

Energy savings in residential and commercial buildings. Various studies have shown that increases in energy efficiency can improve living standards in residential buildings with little or no increase in energy use. Policy must be developed to overcome institutional barriers and constraints in the incentives structure (p. 45). Areas for review include (i) opportunities for improving energy-efficiency incentives for homeowners and occupants through ongoing housing reform initiatives; (ii) heat, gas, and electricity pricing policies for residences and commercial buildings, and the potential use of differential connection charges to encourage energy efficiency; (iii) measures to improve the application and enforcement of energy-efficiency codes for buildings; and (iv) opportunities for introducing innovative financing and cost-sharing schemes involving two or more parties in housing construction, ownership, and occupancy.

Support is needed for development, demonstration, and marketing of new energy-saving construction products, including high-quality, double-paned windows with good-quality weather-strip, hollow bricks, insulation (especially insulated wall paneling), heating system controls, and improved coal stoves. There is a need to integrate more efficient district heating system designs with energy-efficient housing block construction projects.

ALTERNATIVE ENERGY DEVELOPMENT

Expanded use of less carbon-intensive alternatives to coal is a second important objective for the energy sector in China's GHG reduction strategy. To contribute significantly to China's energy economy over the medium and long term, however, greater support for the development of low- or non-carbon energy technologies is urgently needed today. Accordingly, the joint study team recommends that the government establish, with international assistance where required, an aggressive program to accelerate the technological development of alternative energy sources, particularly renewable energy technologies.

The following two conclusions of this study, detailed in Chapters 2 and 3, pose the central problem that must be addressed in China's GHG strategy:

⚬ ...at expanded development, beyond current plans, of the principal alternatives to coal is cost-prohibitive using currently commercial technologies, even when accounting for the high costs of measures to reduce local environmental impacts of coal use. Relying solely on market forces, there is little chance that the dominance of coal in China's energy mix will decline significantly over the medium term. A significant decline in the share of coal use by 2020 based on current technology can be achieved only at enormous financial cost, and therefore cost reduction through technology development is critical.

• Increased reliance on non-coal energy alternatives is the only means to reduce GHG emissions over the long term, short of unacceptable constraints on income growth. The development of alternatives to exponential growth in coal use also will become increasingly essential from both local environmental and logistical points of view. Although energy efficiency is critical over the medium term, only low-or non-carbon fuels can solve the long-term GHG emissions problem.

Large-scale development of coal alternatives over the medium and long term requires sustained policy support and strategically placed investment for technology development and demonstration today. China's energy industry is currently guided by short-term objectives focused on alleviating shortages and adjusting to the new market environment. Institutional responsibilities for developing alternative energy sources are fragmented. The government needs to continue its efforts to establish a well-targeted, clear strategy for the development of cost-effective alternative energy technology.

One of the top priorities in China's GHG reduction strategy, therefore, should be to develop and implement a series of new initiatives to accelerate the development of alternative energy technologies. Primary emphasis should be given to technologies that are most likely to contribute significantly to China's long-term energy supply. The program should focus on research and development, technology transfer from abroad, and technology demonstration and dissemination activities aimed at reducing the costs of alternative energy supply and improving its cost-effectiveness when compared with the use of coal. These activities require support today, including expanded support from the international community and GEF, if alternative energy technologies are to make a significant, commercially viable contribution in the future.

Experts believe that several alternative energy options hold substantial potential for the future if further advances in technology can be made. A few examples include more cost-effective methods for harnessing nuclear power, wind farms based on large-scale generators, advances in solar photovoltaic and thermal-electric technologies, large-scale biomass energy utilization schemes, and new methods for extracting natural gas under difficult geological conditions. Transfer of advanced technology from developed countries will be important in the proposed technology development effort, but China faces an additional challenge to develop very large new supplies of non-coal energy for economic development. Technical leadership in China, therefore, is also necessary.

Several alternative energy technologies exist that are cost-effective compared with coal under a wide variety of conditions. Technical assistance or the transfer and demonstration of approaches or techniques new to China may accelerate development. Two examples include expanded coal-bed methane extraction and use, and further development of sustainable biomass fuel use, such as through the development of high-yield plantations for fuelwood production for direct use or for power generation.

GHG CONTROL IN THE FORESTRY SECTOR

Under the right conditions, afforestation and forestry protection programs can provide significant GHG reduction at low net cost. The afforestation programs that are most important for carbon sequestration are different from those that are most important for other environmental reasons such as erosion control, watershed management, or biological diversity. The objective of carbon sequestration is to amass the most carbon in standing biomass and in the soil at the lowest cost. The following afforestation and forestry management practices have the greatest potential for cost-effective reduction of net carbon emissions in China: (i) the planting of fast-growing, high-yield (FGHY) timber plantations on good quality land; (ii) the establishment of certain multiple-use protection forests; (iii) the planting of high-yield fuelwood plantations on good sites, particularly in southern China; and (iv) improved management and supplemental planting of open forests, also predominantly in southern China. Although fuelwood plantations themselves do not provide significant carbon sequestration, they can significantly reduce carbon emissions by limiting destructive cutting of natural forests and by substituting fuelwood for coal and other fossil fuels in direct use or for power generation.

Given the limited public resources available in China to support forestry development and the country's growing demand for forest products, China should encourage private investment, both domestic and foreign, in commercial timber and fuelwood forestry projects. Policies that can accelerate forestry development in China include (i) improvements in capital markets, (ii) further price reform, and (iii) clarification of land-tenure rights. Improvements in capital markets, including equal access to credit, are needed to allow individuals and firms to borrow for long-term forestry projects. The reform of timber prices in the past several years has helped encourage private sector investment; further reform of log and wood product prices and the removal of international trade barriers would improve allocation of wood products in China. With respect to land rights, national and local governments in China must provide legal assurances to individuals or groups that they can secure the gains of forestry development in the distant future or will be able to transfer those rights to others.

The government can promote more cost-effective development and management of priority afforestation projects through technology transfer, demonstration, and technical assistance. Among the areas of greatest significance are the improvement of (i) growing stock, including genetic improvement, seed supply, and plant propagation systems; (ii) site management, including site identification and classification, fertilization, and weed control; and (iii) stand management, including tree spacing, thinning, and pruning. Human resource capacity building is needed in silviculture and nursery management, forestry research and extension, market analysis, and afforestation model design.

GHG CONTROL IN THE AGRICULTURAL SECTOR

The three primary sources of GHG emissions from China's agricultural sector are rice fields, nitrogen fertilizer, and domesticated animals. The practical, no-regrets options for reducing emissions from domesticated animals have been identified by this study and many have already been adopted in China. China should expand and accelerate programs that increase the efficiency of livestock production while reducing the amount of methane generated per unit of animal product or unit of work. In addition, China should promote research and development of no-regrets options for reducing methane from rice fields and N_2O from fertilizer.

The introduction of improved breeding stock and the expansion of ammoniated straw and feed supplements have had positive financial and social benefits and have reduced methane emissions. One way of expanding these programs is to improve rural financial institutions. Agricultural extension is also needed to popularize these practices and to demonstrate the proper techniques and financial benefits. The ammoniated feed program is of particular significance since it will help reduce the burning of crop residues in the fields. Crop residue burning, in addition to causing local air pollution, contributes non-CO_2 emissions such as methane, N_2O, NOx, and CO to the atmosphere.

INTERNATIONAL ASSISTANCE FOR REDUCING GHG EMISSIONS IN CHINA

The two primary modes of international assistance to China for GHG reduction are (i) non-targeted assistance through conventional programs, and (ii) targeted global environmental assistance, such as through the Global Environment Facility (GEF). In addition, the Chinese government should seek to maximize the efficiency of technologies and processes that are brought to China. Private sector foreign investment will play a large role in the modernization of China's capital stock over the next several decades, and this modernization will have an enormous impact on energy consumption and GHG emissions well into the future. While not addressed in this study, further research should be undertaken to ensure that private sector investment is as efficient as possible.

Conventional development assistance programs

Individual countries and the international community have a strong interest in helping China reduce GHG emissions. China is receiving various types of conventional international assistance which have a direct impact on GHG reduction. Although climate change mitigation is not the primary—or even an intended—objective of most of this assistance, these efforts are among the most important measures that can be taken by the international community to reduce GHG emissions in China.

POLICY REFORM

International development organizations have played a small but important role in supporting the reform of China's economy over the past ten years and such efforts will continue to be critical for effi-

cient resource allocation. Through their lending and technical assistance programs, multilateral agencies have advanced the policy dialogue on price reform, capital market development, and enterprise management and ownership reforms. Energy price reform has been an important part of the policy discussion, with international agencies providing policy studies and recommendations for energy price reform programs in specific sectors. Perhaps most notable have been the efforts in the reform of natural gas and electric power tariffs. Continuing support of China's capital market and enterprise management and ownership reform programs, which are critical for improving economic and energy efficiency, will be key components of multilateral development assistance to China in the 1990s.

SUPPORT FOR RESTRUCTURING AND NO-REGRETS PROJECTS

International development institutions contribute to China's GHG reduction strategy by lending for projects that improve resource allocation in general and energy use in particular. Through their lending programs to China's industrial and energy sectors, international agencies can contribute to the introduction of energy-efficient technologies, the promotion of restructuring programs and achievement of scale economies, and the adoption of modern management methods.

Many of the priority projects for GHG reduction in China outlined on the previous pages are being supported through bilateral and multilateral assistance. For instance, multilateral lending agencies have recently approved investment projects in China for industrial energy conservation, efficient power development, timber and protective forestry management and afforestation, and improved animal feed. International development agencies have also provided support to China for pilot initiatives in alternative energy, coal-bed methane extraction and use, and improved coal utilization. While these projects have not been justified on the basis of climate change mitigation, their contribution to GHG reduction is significant and could be further documented in the course of project preparation.

Role of the Global Environment Facility

The joint study team recommends that GEF resources be used in China to promote and accelerate the priority projects for GHG reduction outlined above.

The broad objective of the GEF for climate change mitigation is to help developing countries reduce or minimize GHG emissions subject to their development goals. While the specific criteria for GEF II have not been finalized, the following principles are likely to guide investment and technical assistance projects:

• GEF funds should be used specifically to advance global environmental objectives and not as another source of development funding.

• GEF should fund projects that are national priorities of the recipient country and that are part of the country's climate change convention strategy.

• GEF resources should be used to achieve the largest reductions in GHG emissions per unit of GEF support. This means that GEF support should i) be used for projects that are replicable; that is, able to proceed without government or other concessional financing after GEF support has been exhausted; and ii) leverage other sources of funding for GHG reduction, particularly domestic private funds.

EXPECTED CRITERIA FOR GEF SUPPORT

Based on the study team's experience with the GEF to date, it is expected that the project areas for GEF support will generally be required to meet each of the following criteria:

• The project areas targeted for assistance must demonstrate the potential for major reductions in GHG emissions as measured in terms of tons of carbon-equivalent per year. The reduction potential should be compared with current and projected GHG emissions, estimates of which are given in Chapters 1 and 2.

• Assistance should support investment and technical assistance projects that are currently cost-effective in terms of unit costs for GHG reduction or that show the clear potential for being cost-effective in the future. Although difficult to quantify precisely, the most cost-effective GEF projects are likely to be those that address capacity building, institutional strengthening, imperfect information, or other market and non-market barriers to no-regrets GHG reduction projects.

• A strong case must be made that the project in question could not proceed without GEF support and that the constraints are other than general investment capital shortages or policy distortions (such as subsidies or price controls). In general, this means that projects normally eligible for assistance through commercial or development banks would not be funded by the GEF.

The priority projects for GHG reduction in China outlined in this chapter meet the first two criteria, but not necessarily the third.

The following areas for GEF support are consistent with the principles and criteria outlined above.

OVERCOMING MARKET AND NON-MARKET BARRIERS

GEF resources should be used to overcome non-policy barriers to projects that yield significant benefits to the national economy. The case study analyses of Chapter 3 have shown that even where the financial and economic returns are good and where financing is not a constraint, some projects are not proceeding as fast as they should. While many of these barriers exist in well-developed market economies, some are more pronounced in China because the economy is in transition. GEF resources could help overcome the following barriers: i) imperfect or asymmetrical information, ii) technical and market risk, and iii) high transactions costs.

Imperfect information. Information that serves a purely public good, such as climate change mitigation, is unlikely to be provided by the private sector; thus, projects that depend on such information may not be undertaken. Public education campaigns, technology dissemination, extension services, and human resources may be needed to overcome information barriers.

Public education. Public education campaigns can address the lack of information which is often a barrier to the adoption of new equipment or ways of doing business. Public campaigns, which have been used extensively in China, can and are being used by central and local governments to promote energy conservation and afforestation. Consumer labeling, such as standardized data on the life-cycle costs of major energy-consuming equipment, enables consumers to make more informed decisions when purchasing capital equipment.

Dissemination of technology and techniques. Once projects have been proven successful and commercially viable, the government can assist in promoting the technology throughout the country. Extension, particularly in the agricultural sector, has been one of China's institutional strengths. In addition to agricultural programs, expanded efforts are needed for dissemination of information on no-regrets projects in energy efficiency, forestry, and alternative energy.

Training. Without the concurrent development of human capital, many of the options for reducing GHG emissions will not succeed. For instance, much of the energy-efficiency savings associated with new equipment cannot be realized unless operators are trained to use and maintain the equipment. Local service centers can be set up to provide enterprises with assistance and training on energy use, energy audits, technical troubleshooting, and investment appraisal and post-investment monitoring. Training is especially needed in financial and economic analysis to demonstrate the returns to energy efficiency and other types of investments.

Imperfect markets for risk. The GEF can help overcome excessive risk associated with the acquisition or demonstration of new technologies that are important for GHG reduction. Even in countries with well-developed capital markets, technical and market risk can hinder the development of superior technologies. In developing countries such as China, the risk associated with adopting new technologies or processes is even greater. Faced with the choice of investing in a proven process that increases output or a relatively unknown technology that saves energy, many enterprises opt for the former. When such "risk premiums" are large, or where the benefits for an individual firm are relatively small, there is almost no chance that the superior technology will be adopted until the risk can be reduced or shared.

Technology transfer. For enterprises to invest in new technologies or processes, there must be private returns to that investment. In practice, this often means that the technologies are held in strict confidence by the developer or that investment to develop a technology is not made because the nature of the technology makes it nearly impossible to appropriate private returns. When the objective is to maximize GHG reductions, it is in society's interest to have leading technologies and processes in the public domain. The GEF could help China acquire internationally advanced technologies having broad applicability and major GHG reduction benefits, but which would otherwise have insufficient returns to domestic developers. Industrial technologies related to energy production and consumption are key areas for technology transfer. A wide gap exists between Chinese and international technologies in industrial boilers, coal gasification, and gas recovery systems from coal

mines. While industrial technologies and processes are the most obvious examples, the import of technologies in agriculture and forestry, such as improved cattle species or advanced silviculture techniques, would also benefit GHG reduction.

Demonstration projects. To reduce the risk associated with new technologies and processes that have the potential for cost-effective GHG reduction, the GEF could support selected demonstration projects. In addition to overcoming technical risk, demonstration projects allow information to be disseminated on the financial aspects of adopting the new technology. In return for subsidies, enterprises undertaking demonstration projects should be required to provide detailed information on both the technical and financial aspects of the project. Seed monies could be provided to enterprises to undertake demonstration projects with commercial returns, and proceeds could be used to promote additional high-priority projects.

Transaction costs. When there are many consumers, the costs of acquiring information or negotiating with producers become prohibitive. For instance, it is nearly impossible for the thousands of consumers of steam traps, insulated pipe, water pumps, or electric motors to negotiate with producers to demand more efficient products. Without consumer product labeling or efficiency standards it may be too costly for individual consumers to acquire information or to inform producers how much they are willing to pay for more energy-efficient equipment.

Standards. In parallel with product labeling, China should expand and improve its system of energy-efficiency standards. Such standards should be adopted for key consumer goods, such as refrigerators and air conditioners, and for major energy-consuming equipment, such as electrical motors and industrial boilers. Codes for building materials (glass, bricks, wallboard) and for building construction should also be selectively adopted.

Distributors of World Bank Publications

Prices and credit terms vary from country to country. Consult your local distributor before placing an order.

ALBANIA
Adrion Ltd.
Perlat Rexhepi Str.
Pall. 9, Shk. 1, Ap. 4
Tirana
Tel: (42) 274 19; 221 72
Fax: (42) 274 19

ARGENTINA
Oficina del Libro Internacional
Av. Cordoba 1877
1120 Buenos Aires
Tel: (1) 815-8156
Fax: (1) 815-8354

AUSTRALIA, FIJI, PAPUA NEW GUINEA, SOLOMON ISLANDS, VANUATU, AND WESTERN SAMOA
D.A. Information Services
648 Whitehorse Road
Mitcham 3132
Victoria
Tel: (61) 3 9210 7777
Fax: (61) 3 9210 7788
URL: http://www.dadirect.com.au

AUSTRIA
Gerold and Co.
Graben 31
A-1011 Wien
Tel: (1) 533-50-14-0
Fax: (1) 512-47-31-29

BANGLADESH
Micro Industries Development Assistance Society (MIDAS)
House 5, Road 16
Dhanmondi R/Area
Dhaka 1209
Tel: (2) 326427
Fax: (2) 811188

BELGIUM
Jean De Lannoy
Av. du Roi 202
1060 Brussels
Tel: (2) 538-5169
Fax: (2) 538-0841

BRAZIL
Publicações Tecnicas Internacionais Ltda.
Rua Peixoto Gomide, 209
01409 Sao Paulo, SP.
Tel: (11) 259-6644
Fax: (11) 258-6990

CANADA
Renouf Publishing Co. Ltd.
1294 Algoma Road
Ottawa, Ontario K1B 3W8
Tel: 613-741-4333
Fax: 613-741-5439

CHINA
China Financ... & Economic Publishing House
8, Da Fo Si Dong Jie
Beijing
Tel: (1) 333-8257
Fax: (1) 401-7365

COLOMBIA
Infoenlace Ltda.
Apartado Aereo 34270
Bogotá D.E.
Tel: (1) 285-2798
Fax: (1) 285-2798

COTE D'IVOIRE
Centre d'Edition et de Diffusion Africaines (CEDA)
04 B.P. 541
Abidjan 04 Plateau
Tel: 225-24-6510
Fax: 225-25-0567

CYPRUS
Center of Applied Research
Cyprus College
6, Diogenes Street, Engomi
P.O. Box 2006
Nicosia
Tel: 244-1730
Fax: 246-2051

CZECH REPUBLIC
National Information Center
prodejna, Konviktska 5
CS – 113 57 Prague 1
Tel: (2) 2422-9433
Fax: (2) 2422-1484
URL: http://www.nis.cz/

DENMARK
SamfundsLitteratur
Rosenoerns Allé 11
DK-1970 Fredeniksberg C
Tel: (31)-351942
Fax: (31)-357822

EGYPT, ARAB REPUBLIC OF
Al Ahram
Al Galaa Street
Cairo
Tel: (2) 578-6083
Fax: (2) 578-6833

The Middle East Observer
41, Sherif Street
Cairo
Tel: (2) 393-9732
Fax: (2) 393-9732

FINLAND
Akateeminen Kirjakauppa
P.O. Box 23
FIN-00371 Helsinki
Tel: (0) 121411
Fax: (0) 121-4441
URL: http://booknet.cultnet.fi/aka/

FRANCE
World Bank Publications
66, avenue d'Iéna
75116 Paris
Tel: (1) 40-69-30-56/57
Fax: (1) 40-69-30-68

GERMANY
UNO-Verlag
Poppelsdorfer Allee 55
53115 Bonn
Tel: (228) 212940
Fax: (228) 217492

GREECE
Papasotiriou S.A.
35, Stournara Str.
106 82 Athens
Tel: (1) 364-1826
Fax: (1) 364-8254

HONG KONG, MACAO
Asia 2000 Ltd.
Sales & Circulation Department
Seabird House, unit 1101-02
22-28 Wyndham Street, Central
Hong Kong
Tel: 852 2530-1409
Fax: 852 2526-1107
URL: http://www.sales@asia2000.com.hk

HUNGARY
Foundation for Market Economy
Dombovari Ut 17-19
H-1117 Budapest
Tel: 36 1 204 2951 or 36 1 204 2948
Fax: 36 1 204 2953

INDIA
Allied Publishers Ltd.
751 Mount Road
Madras - 600 002
Tel: (44) 852-3938
Fax: (44) 852-0649

INDONESIA
Pt. Indira Limited
Jalan Borobudur 20
P.O. Box 181
Jakarta 10320
Tel: (21) 390-4290
Fax: (21) 421-4289

IRAN
Kowkab Publishers
P.O. Box 19575-511
Tehran
Tel: (21) 258-3723
Fax: 98 (21) 258-3723

IRELAND
Government Supplies Agency
Oifig an tSoláthair
4-5 Harcourt Road
Dublin 2
Tel: (1) 461-3111
Fax: (1) 475-2670

ISRAEL
Yozmot Literature Ltd.
P.O. Box 56055
Tel Aviv 61560
Tel: (3) 5285-397
Fax: (3) 5285-397

R.O.Y. International
PO Box 13056
Tel Aviv 61130
Tel: (3) 5461423
Fax: (3) 5461442

Palestinian Authority/Middle East
Index Information Services
P.O.B. 19502 Jerusalem
Tel: (2) 271219

ITALY
Licosa Commissionaria Sansoni SPA
Via Duca Di Calabria, 1/1
Casella Postale 552
50125 Firenze
Tel: (55) 645-415
Fax: (55) 641-257

JAMAICA
Ian Randle Publishers Ltd.
206 Old Hope Road
Kingston 6
Tel: 809-927-2085
Fax: 809-977-0243

JAPAN
Eastern Book Service
Hongo 3-Chome,
Bunkyo-ku 113
Tokyo
Tel: (03) 3818-0861
Fax: (03) 3818-0864
URL: http://www.bekkoame.or.jp/~svt-ebs

KENYA
Africa Book Service (E.A.) Ltd.
Quaran House, Mfangano Street
P.O. Box 45245
Nairobi
Tel: (2) 23641
Fax: (2) 330272

KOREA, REPUBLIC OF
Daejon Trading Co. Ltd.
P.O. Box 34
Yeoeida
Seoul
Tel: (2) 785-1631/4
Fax: (2) 784-0315

MALAYSIA
University of Malaya Cooperative Bookshop, Limited
P.O. Box 1127
Jalan Pantai Baru
59700 Kuala Lumpur
Tel: (3) 756-5000
Fax: (3) 755-4424

MEXICO
INFOTEC
Apartado Postal 22-860
14060 Tlalpan,
Mexico D.F.
Tel: (5) 606-0011
Fax: (5) 606-0386

NETHERLANDS
De Lindeboom/InOr-Publikaties
P.O. Box 202
7480 AE Haaksbergen
Tel: (53) 574-0004
Fax: (53) 572-9296

NEW ZEALAND
EBSCO NZ Ltd.
Private Mail Bag 99914
New Market
Auckland
Tel: (9) 524-8119
Fax: (9) 524-8067

NIGERIA
University Press Limited
Three Crowns Building Jericho
Private Mail Bag 5095
Ibadan
Tel: (22) 41-1356
Fax: (22) 41-2056

NORWAY
Narvesen Information Center
Book Department
P.O. Box 6125 Etterstad
N-0602 Oslo 6
Tel: (22) 57-3300
Fax: (22) 68-1901

PAKISTAN
Mirza Book Agency
65, Shahrah-e-Quaid-e-Azam
P.O. Box No. 729
Lahore 54000
Tel: (42) 7353601
Fax: (42) 7585283

PERU
Editorial Desarrollo SA
Apartado 3824
Lima 1
Tel: (14) 285380
Fax: (14) 286628

PHILIPPINES
International Booksource Center Inc.
Suite 720, Cityland 10
Condominium Tower 2
H.V dela Costa, corner
Valero St.
Makati, Metro Manila
Tel: (2) 817-9676
Fax: (2) 817-1741

POLAND
International Publishing Service
Ul. Piekna 31/37
00-577 Warzawa
Tel: (2) 628-6089
Fax: (2) 621-7255

PORTUGAL
Livraria Portugal
Rua Do Carmo 70-74
1200 Lisbon
Tel: (1) 347-4982
Fax: (1) 347-0264

ROMANIA
Compani De Librarii Bucuresti S.A.
Str. Lipscani no. 26, sector 3
Bucharest
Tel: (1) 613 9645
Fax: (1) 312 4000

RUSSIAN FEDERATION
Isdatelstvo <Ves Mir>
9a, Kolpachniy Pereulok
Moscow 101831
Tel: (95) 917 87 49
Fax: (95) 917 92 59

SAUDI ARABIA, QATAR
Jarir Book Store
P.O. Box 3196
Riyadh 11471
Tel: (1) 477-3140
Fax: (1) 477-2940

SINGAPORE, TAIWAN, MYANMAR, BRUNEI
Asahgate Publishing Asia Pacific Pte. Ltd.
41 Kallang Pudding Road #04-03
Golden Wheel Building
Singapore 349316
Tel: (65) 741-5166
Fax: (65) 742-9356
e-mail: ashgate@asianconnect.com

SLOVAK REPUBLIC
Slovart G.T.G. Ltd.
Krupinska 4
PO Box 152
852 99 Bratislava 5
Tel: (7) 839472
Fax: (7) 839485

SOUTH AFRICA, BOTSWANA
For single titles:
Oxford University Press Southern Africa
P.O. Box 1141
Cape Town 8000
Tel: (21) 45-7266
Fax: (21) 45-7265

For subscription orders:
International Subscription Service
P.O. Box 41095
Craighall
Johannesburg 2024
Tel: (11) 880-1448
Fax: (11) 880-6248

SPAIN
Mundi-Prensa Libros, S.A.
Castello 37
28001 Madrid
Tel: (1) 431-3399
Fax: (1) 575-3998
http://www.tsai.es/mprensa

Mundi-Prensa Barcelona
Consell de Cent, 391
09009 Barcelona
Tel: (3) 488-3009
Fax: (3) 487-7659

SRI LANKA, THE MALDIVES
Lake House Bookshop
100, Sir Chittampalam A. Gardiner Mawatha
Colombo 2
Tel: (1) 32105
Fax: (1) 432104

SWEDEN
Fritzes Customer Service
Regeringsgaton 12
S-106 47 Stockholm
Tel: (8) 690 90 90
Fax: (8) 21 47 77

Wennergren-Williams AB
P. O. Box 1305
S-171 25 Solna
Tel: (8) 705-97-50
Fax: (8) 27-00-71

SWITZERLAND
Librairie Payot
Service Institutionnel
Côtes-de-Montbenon 30
1002 Lausanne
Tel: (021)-341-3229
Fax: (021)-341-3235

Van Diermen Editions Techniq
Ch. de Lacuez 41
CH1807 Blonay
Tel: (021) 943 2673
Fax: (021) 943 3605

TANZANIA
Oxford University Press
Maktaba Street
PO Box 5299
Dar es Salaam
Tel: (51) 29209
Fax: (51) 46822

THAILAND
Central Books Distribution
306 Silom Road
Bangkok
Tel: (2) 235-5400
Fax: (2) 237-8321

TRINIDAD & TOBAGO, JAM.
Systematics Studies Unit
#9 Watts Street
Curepe
Trinidad, West Indies
Tel: 809-662-5654
Fax: 809-662-5654

UGANDA
Gustro Ltd.
Madhvani Building
PO Box 9997
Plot 16/4 Jinja Rd.
Kampala
Tel/Fax: (41) 254763

UNITED KINGDOM
Microinfo Ltd.
P.O. Box 3
Alton, Hampshire GU34 2PG
England
Tel: (1420) 86848
Fax: (1420) 89889

ZAMBIA
University Bookshop
Great East Road Campus
P.O. Box 32379
Lusaka

ZIMBABWE
Longman Zimbabwe (Pte.)Ltd.
Toune Road, Ardbennie
P.O. Box ST125
Southerton
Harare
Tel: (4) 6216617
Fax: (4) 621670